Elias Gross

An elementary treatise on kinematics and kinetics

Elias Gross

An elementary treatise on kinematics and kinetics

ISBN/EAN: 9783337277888

Printed in Europe, USA, Canada, Australia, Japan

Cover: Foto ©berggeist007 / pixelio.de

More available books at **www.hansebooks.com**

Rivingtons' Mathematical Series

KINEMATICS AND KINETICS

AN ELEMENTARY TREATISE

ON

KINEMATICS AND KINETI

BY

E. J. GROSS, M.A.

FELLOW OF GONVILLE AND CAIUS COLLEGE, CAMBRIDGE, AND SECRETARY TO THE
OXFORD AND CAMBRIDGE SCHOOLS EXAMINATION BOARD

RIVINGTONS

London, Oxford, and Cambridge

1876

RIVINGTONS

London . . . . Waterloo Place
Oxford . . . . High Street
Cambridge . . . . Trinity Street

[B—58]

PREFACE.

THIS Treatise is intended to contain as much as is required, under the head of Dynamics, of Candidates for Honours in the First Three Days of the Mathematical Tripos. I hope that it will also be of use to Students in their preparation for other Examinations, where questions are set which may be treated without Analytical Geometry and the Differential Calculus.

A beginner, who wishes to become acquainted with the principles of Dynamics before advancing far in the Kinematical portion of the book, will find that Chapters VII. and VIII. may be read immediately after Chapter I.

My thanks are due to Mr. Hamblin Smith for having kindly examined most of the proof sheets as they passed through the press.

I shall be very grateful for any corrections, or suggestions for the improvement of the work, which may be sent me by any one using it.

E. J. GROSS.

GONVILLE AND CAIUS COLLEGE,
November 23, 1875.

CONTENTS.

KINEMATICS.

KINETICS.

ERRATA.

Page 84, line 2 from top, *for* 6928 *read* 7728.
Page 114, line 7 from top, *for* chord *read* cord.

KINEMATICS.

I.—VELOCITY.

1. When the position of a point is being changed continuously, the point is said to be in *motion*.

2. *Velocity* is the name given to the *rate* of motion of the point, or the *degree of quickness* or *slowness* with which the point is moving, at any instant.

3. By observing a body, such as a train, in motion, we perceive sometimes that it is moving faster at one instant than at another.

Again, we see sometimes, when two bodies are in motion, that one is moving faster than the other.

We can express these facts by saying that the velocity of the train is greater at one instant than at another; and that, in the second case, the velocity of one body is greater than that of the other.

We thus become familiar with the idea of velocities differing from one another in *magnitude*, or *intensity*.

And we see that the velocity of a point is a property, which the point has at each individual instant of its motion, and that the magnitude of this property may be different at different instants.

A

4. If during any interval of time the magnitude of the velocity of a point is the same at every instant, the velocity is said to be *uniform.* If the magnitude at one instant is different from what it is at another, the velocity is said to vary, or to be a *variable* velocity.

5. If a point always moves uniformly, it is easy to see that it will pass over equal distances, or spaces, in equal intervals of time. But the converse of this, viz., that, if equal spaces are passed over in equal times, the velocity is uniform, is not necessarily true. Thus, if the velocity of a train is uniform and of the proper magnitude, it will pass over 30 miles in every hour. But, if it passes over 30 miles in every hour, it does not by any means follow that its velocity is uniform throughout an hour; for at one part of that time it may be moving faster than at another, if only it manage upon the whole to go exactly 30 miles in the hour. If, however, it went ½ mile in every minute, we should feel more confident that it was moving uniformly, and still more so, if it went $\frac{1}{120}$ mile in every second. And if, on dividing the time into equal intervals as small as we pleased, we found that it went an equal space in each interval, we should conclude that the velocity was uniform throughout the whole time. We thus arrive at the following *test* for uniformity of velocity :—

A velocity of a point is said to be uniform when equal spaces are passed over in equal intervals of time, *however small.*

6. Our ideas of the velocities of points are closely connected with those of the spaces, over which the points will go in any specified time.

Thus, in *ordinary language*, we indicate any particular velocity by mentioning the space traversed in some given length of time by a point moving uniformly with that velocity during this time.

For example, we talk of a velocity of 40 miles an hour ; meaning such a velocity that, if a point travel uniformly with it for an hour, the point will pass over 40 miles.

Ex. **A point possesses** a velocity of 20 miles per 75′. How far will it go in an hour ?

$$\text{The point goes } \mathbf{20} \text{ miles in } \mathbf{75'};$$

$$\therefore \text{ it } \quad „ \quad \frac{20}{75} \quad „ \quad 1';$$

$$\therefore \text{ it } \quad „ \quad \frac{20}{75} \times 60 \quad „ \quad 60';$$

$$i.e. \text{ „ } \quad „ \quad 16 \quad „ \quad 1\text{hr}.$$

EXAMPLES.—I.

(1.) A point is travelling with a velocity of 20 feet a minute. How far will it go in an hour ?

(2.) Through how many yards will a point go in half-an-hour, whose velocity is 15 feet per second ?

(3.) A point is travelling with a velocity of 2 miles an hour. (1) How far will it go in a minute ? (2) How many feet will it traverse in 10″ ?

(4.) If a point moves uniformly over 3 feet in a second, by how many miles an hour would you represent this velocity ?

(5.) The velocity of a point is v feet a minute, and in a quarter of an hour it has gone a mile. What is v ?

(6.) The **velocity of a train** is 30 miles an hour. (1) How long will it take to traverse 100 yards ? (2) How many seconds will it take to go 150 feet ?

(7.) A train travels 120 miles in two hours and a half. If it travels uniformly, how many yards does it go in a minute ?

7. Further, we can compare velocities by comparing the spaces, over which points possessing these velocities will go in any the same time.

Thus, by one velocity being double another, we should understand that, if for any particular time, say a minute, two points were to move uniformly, one with the first velocity and the other with the second, then the first point would go in the minute twice as far as the second.

Again, if the first velocity were v times as great as the second, the first point would go, in any particular time, v times as far as the second. In this case we should say that the first velocity *contained* the second v times.

Ex. A velocity of 2 yards a second is half that of 720 feet a minute.

For a point having the first velocity will go 2 yards a second,

i.e. 6 feet   ,,   ;

and   ,,   other ,, will go 720 feet a minute,

i.e. 12 feet a second.

Hence the number of feet traversed with the first velocity in a second is half the number traversed with the other in the same time. Therefore the first velocity is half the other.

EXAMPLES.—II.

(1.) Show that a velocity of 60 miles an hour is double that of 44 feet a second.

(2.) How many yards an hour must a velocity be in order to be triple one of 2 miles a minute?

(3.) How long must a point take to go 3 miles, (1) in order that its velocity may be 5 times that of 8 feet a second? (2) If its velocity is $\frac{1}{2}$ of that of 20 yards a minute?

(4.) Find the ratio between two velocities, one being 70 miles an hour, and the other two yards a second.

(5.) A velocity of 3 yards a second is v times one of 70 feet a minute. What is v?

(6.) Compare the velocities of 20 miles an hour and 4 yards a minute.

(7.) Compare the velocities of two points moving uniformly, one over 20 miles in an hour, and the other 2 feet in a quarter of a second.

(8.) One body moves over 30 yards in 7 minutes, and the other over 60 feet in 25 seconds. If their velocities are uniform, compare them.

(9.) How many times does the velocity of 300 yards per 11 minutes contain the velocity of 70 feet per 3 seconds?

(10.) The velocity 20 miles an hour contains the velocity 30 feet a second v times. What is v?

(11.) A man 6 feet high walks in a straight line at the rate of 4 miles an hour away from a street lamp, the height of which is 10 feet. Supposing the man to start from the lamp-post, find the rate at which the end of his shadow travels, and also the rate at which his shadow lengthens.

8. We explained in Art. **6** how a velocity may be indicated in ordinary language. We will now state how it may be represented by a number, on which the operations of Algebra may be performed.

In *Algebra* the magnitude of a velocity, like everything else, is represented by its measure (Alg. Pt. I. Art. 33). That is to say, we fix upon some definite velocity, with which we are familiar, as our standard, and represent any particular velocity by the number of times it contains this standard, this number being called the *measure* of the particular velocity.

Instead of the phrase, "the velocity whose measure is v," we often write the shorter one "the velocity v." So that, when we put a letter for a velocity, the student must remember that it only indicates the *number* of times that the velocity contains the standard.

The velocity which we fix upon as our standard is that, with which a point, if it moves uniformly, will pass over a unit of space in a unit of time.

9. PROP. *The measure of a velocity is equal to the measure of the space traversed in a unit of time by a point moving uniformly with the velocity.*

Let v denote the measure of the velocity, *i.e.* let it contain the standard v times. Therefore, Art. **7**, a point, travelling with this velocity, would go v times as far in a unit of time as if it were travelling with the standard.

Now with the standard the point would go one unit of space in a unit of time, and therefore with the velocity v it will go v units of space in a unit of time, or, in other words, v is the measure of the space it traverses in a unit of time.   Q.E.D.

Cor. Suppose a point travels uniformly for a time t with a velocity v. By the Prop. it traverses a space v in each unit of time, hence in t units of time it traverses a space vt.

And further, if at the beginning of the time its distance from a fixed point in its line of motion is a, its distance at the end is $a \pm vt$, according as it has moved from, or towards, the fixed point.

Let s denote the space traversed in time t. Then, if it move uniformly with velocity v, we have $s = vt$, or $\dfrac{s}{t} = v$. Conversely, if the fraction $\dfrac{s}{t}$ is always the same for all values of t (*i.e.* if $s \propto t$), we conclude that the velocity is constant or *uniform*.

This includes the test of Art. 5.

Note.—The standard velocity is called the unit of velocity, or the unit velocity.

10. *Ex.* 1. If the unit of time be a minute and a foot the unit of space, what is the measure of the velocity 40 miles an hour?

A point having this velocity goes 40 miles in an hour.

$$\text{i.e. } 40 \times 1760 \times 3 \text{ feet} \qquad \text{,,}$$
$$\text{i.e. } \frac{40 \times 1760 \times 3}{60} \qquad \text{,, in a minute,}$$
$$\text{i.e. } 3520 \qquad \text{,,} \qquad \text{,,}$$

Hence, the measure of the space passed over in a unit of time being 3520, the measure of the velocity is also 3520.

If the unit of space had been 10 feet, the measure of the distance traversed in a unit of time would have been 352, and therefore the measure of velocity 352 also.

Ex. 2. How far will a point having a velocity 3 go in 4 units of time?

Here $v = 3$, $t = 4$; $\therefore$ the formula $s = vt$ shows that 12 units of space would be passed over in 4 units of time.

EXAMPLES.—III.

(1.) **What is** the measure of the velocity 70 feet a second when a foot a second is the unit velocity ?

(2.) A yard per 7 seconds is the standard velocity. What is the measure of the velocity 15 yards per 7 seconds?

(3.) **What will be** the measure of the velocity 16 miles an hour (1) when a foot a second is the standard; (2) when 4 **feet** per minute is the standard ?

(4.) How many yards **an hour must** a point traverse, in order **that** the measure of its **velocity may be** $\frac{25}{9}$, when a foot a second is **the** unit velocity ?

(5.) **How many minutes will a** body take to go a mile with a **velocity whose measure is** 5 ; the standard being the velocity of **25** feet per 3′ ?

(6.) A distance of 5 yards is the unit of space, and **an interval of** 3′ is the unit **of** time. How far will a body go in half an **hour** with a velocity 7 ?

(7.) **If the** velocity of 30 feet a second be represented **by** 5, what will be the measure of the velocity 7 yards per 2′ ?

(8.) Two points start from the same position and move in opposite directions with velocities of 5 feet a minute and 10 feet per 3 seconds. How far apart will they be at the end of 5 minutes ?

(9.) Two points start from the same point and move in perpendicular directions, one with a velocity 5 yards a second, and the other with a velocity of 10 feet a **minute.** How **far apart** will they be, (1) at the end of 5′, **when** they start simultaneously ; (2) at the **end of** 10′ **from** the starting of the last one, when the first starts 3′ before **the** second ?

(10.) Two points move along two lines containing an angle of 60°. One point **moves with** the velocity 30 feet a second, and the other **with** the velocity 20 feet per 2″, and they start simultaneously from **the** point **of intersection of** the lines. How far apart will they be at the end of 2′ ?

(11.) A particle whose motion is uniform, is at the end of the day a mile distant from its position at the commencement. Find its velocity, taking 11 yards as the unit of space, 9 minutes as the unit **of time, and a** day equal to 24 hours.

11. The direction in which a point is moving is called the direction of its velocity.

In order to determine completely any velocity, we must determine both its magnitude and its direction.

12. It is to be observed that we can prefix the signs $+$ and $-$ before the measures of velocities, to indicate contrariety of direction, as in Trigonometry, etc.

Thus if $+v$ (or v) indicate the velocity of a point moving to the right, $-v$ will indicate the velocity of a point moving, at an equal rate, to the left.

13. We can represent velocities by straight lines.

For we can draw a straight line

1° in any direction, and thus we can represent the direction of any velocity;

2° so as to contain as many units of length as the velocity contains units of velocity, and thus we can represent the magnitude of the velocity.

Hence, Art. **9,** the straight line will represent also the space which the point would traverse, if it moved uniformly for a unit of time with the velocity represented. And, conversely, if it moves uniformly for a unit of time with any particular velocity, the space traversed will represent that velocity.

EXAMPLES.—IV.

(1.) A velocity of 5 miles an hour is represented by a line 10 inches long. What length of line will represent a velocity of 9 miles an hour?

(2.) A velocity 3 is represented by a line 5 inches long. What line will represent a velocity 7?

(3.) If a velocity of 5 miles an hour to the north is represented by 4, what would represent a velocity of 5 feet a minute to the south?

(4.) Given that a certain line containing 11 inches represents a velocity of 3 miles an hour to the east. How would you represent a velocity of 100 yards a minute to the north-east?

14. When a particular velocity has been taken as our standard, every velocity will have its own certain measure.

If now we change our standard, the measure of each velocity must also be changed.

For instance, if for any reason we took as our standard a velocity double of the previous standard, the measure of every velocity would be half what it was previously, for a velocity which contains the old unit v times would only contain the new unit $\frac{v}{2}$ times.

We have defined our standard velocity with reference to the units of space and time. If, therefore, we change those units, we must, generally, take a new velocity for our standard ; and consequently the measure of every velocity will then be altered.

It is the object of the following proposition to find the change produced in the measure of a velocity by any given change in the units of space and time.

We will first illustrate the method of proof on a particular case, and then apply it to the general proposition.

15. A certain velocity has 5 for its measure when a foot and a second are units of space and time. What will be its measure when a yard and a minute are units ?

Now 5 feet are equal to $\frac{5}{3}$ yards.

With the given velocity a point will traverse (Prop. Art. **9**)

5 feet   in one second,

i.e. $\frac{5}{3}$ yards in one second ;

$\therefore 60 \cdot \frac{5}{3}$ yards in one minute ;

$\therefore$, by Prop. Art. **9**, $60 \cdot \frac{5}{3}$ ($=100$) is the new measure of the velocity required.

16. Prop. *Given the measure of any velocity with certain (old) units of space and time, to find the* **measure of the** *same velocity with any other (new) units.*

Let v be the measure of the velocity with the old **units.**

Let a and b denote the number of times, respectively, which the new units of space and time contain the old.

Hence v old units of space are equal to $\dfrac{v}{a}$ new units.

With the given **velocity a point will traverse (Prop. Art. 9),**
v old units of space in one old unit of time,

$i.e.\ \dfrac{v}{a}$ new units of space in one old unit of time ;

$\therefore b \cdot \dfrac{v}{a}$ new units of space in one new unit of time ;

$\therefore$, by **Prop. Art. 9,** $\dfrac{bv}{a}$ is the new measure (v') of the velocity required. **Hence if we** know the values of any 3 of the symbols v, v', a, b, **we can** find the value of the fourth from the equation $v' = \dfrac{bv}{a}$.

17. We will now apply **this** formula to the solution of examples.

Ex. 1. A point is moving with the velocity of 5 feet a minute. What will **be** the measure of this velocity when a yard and **a second** are the units of space and time ?

1°. The measure is 5 if **we take** a foot and a minute as units. **(Art. 9.)**

2°. The new **units,** viz., a yard and a second, are respectively 3 times and $\frac{1}{60}$ of the old units.

Thus we can put $v=5$, $a=3$, $b=\dfrac{1}{60}$;

$\therefore$ the new measure required $= \dfrac{\dfrac{1}{60} \cdot 5}{3} = \dfrac{1}{36}$

Ex. 2. A velocity of 4 feet per second is the unit of velocity, and 5′ is the unit of time. What is the unit of space?

Let a feet be the unit of space.

Consider this velocity of 4 feet per second.

Its measure would be 4, if a foot and a second were the units; hence, taking these as our old units, we can put $v=4$.

Again its measure is 1, if a feet and 5′ are the units, since it is then the standard, and 5′ = 300″.

Hence we can put $v=4$, $v'=1$, $b=300$; ∴ $1=\dfrac{300.4}{a}$;

∴ $a=1200$.

Therefore a length of 1200 feet is our unit of space.

EXAMPLES.—V.

(1.) The measure of a certain velocity is 7, when 5 feet and 3″ are the units of space and time. What will be its measure when 2 inches and 4′ are the units?

(2.) The velocity of 7 feet per 3″ is the unit of velocity, and the unit of space contains 3 yards. What is the unit of time?

(3.) What is the unit of space, if 5″ is the unit of time, and the unit velocity is 4 feet a minute?

(4.) A point traverses 29 feet in 3″ with the unit velocity. What is the unit of time, if 4 feet is the unit of space?

(5.) What is the unit of time when the velocity of 6 feet a second is represented by 4, and the unit of space is 7 feet?

(6.) Show that the standard velocity varies directly as the unit of space, and inversely as the unit of time.

(7.) If a velocity of 6 miles an hour be the unit of velocity, what must be the unit of time that 11 yards may be the unit of space?

(8.) If a mile per minute were the unit of velocity, and a yard the unit of space, find the unit of time.

(9.) A body moves uniformly through $(m+n)$ feet in $(m-n)″$, the units of space and time being a foot and second; and the numerical representation of its velocity is nine times what it would have been, had it moved through $(m-n)$ feet in $(m+n)″$, and the units of space and time been a yard and 27″. Show that $m:n=5:4$.

(10.) If a shilling be the unit of money, £1000 a year the unit of income, and an inch per minute the unit of velocity, find the unit of space. It being understood that the unit of income is an income of a unit of money in a unit of time.

18. If a traveller in a railway train in motion walks from one window of his carriage to the other, he may be considered as having, at every instant, two velocities; one the same as that of the train, and one independent of the train across the carriage. We thus obtain an idea of how velocities in different directions may coexist in a point.

The traveller's actual velocity will be intermediate in direction between these two coexisting velocities; and it is the object of the proposition of Art. **20** to determine the actual velocity at any instant for this and all similar cases.

19. DEF. When two or more velocities coexist in a point, its actual velocity is called their *resultant;* and each of the coexisting velocities is called a *component* of this resultant.

20. PROP. The Parallelogram of Velocities. *If two straight lines AB, AC, represent two co-existing velocities, u and v, of a point, and the parallelogram ABDC be completed,* **then the** *diagonal* **AD** *will represent the resultant velocity.*

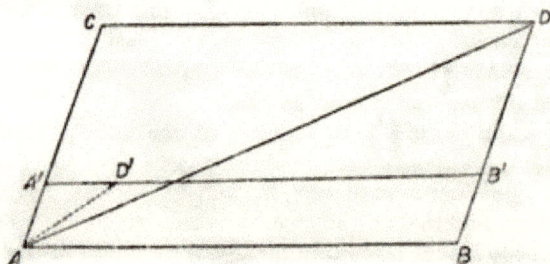

FIG. 1.

Since *AB, AC* represent the velocities; if *v* did not exist and the point retained the velocity *u* for a unit of time, it would pass over AB (Art. 13); and if u did not exist, and it retained v for a unit of time, it would pass over AC. We have to show that if it retain both velocities for a unit of time, it will move uniformly along AD and will pass over AD in the unit of time.

We can represent the coexistence of the velocities by supposing the point to move along AB with the velocity u, whilst AB moves parallel to itself, with the end A along AC, with the velocity v.

1°. The point moves along AD.

After any time, t, let AB be in the position $A'B'$, and the point in the position D'.

Then $A'D' = ut$, $AA' = vt$;

$$\therefore AA' : A'D' = v : u = AC : CD,$$

and $\angle AA'D' = \angle ACD$;

$\therefore$ the triangles $AA'D'$, and ACD are equiangular;

$\therefore \angle A'AD' = \angle CAD$; *i.e.* AD' coincides in direction with AD.

And this being so for all values of t, the point must travel along AD.

2°. It travels *uniformly* along AD.

For $AD' : AD = AA' : AC = vt : v = t : 1$;

$\therefore$ the distance traversed by the point in any time varies as that time ;

$\therefore$ the velocity along AD is uniform. (Art. 9. Cor.)

3°. It will pass over AD in the unit of time.

For at the end of the unit of time AB has arrived at CD, and the point has moved over AB, *i.e.* it has arrived at D.

Hence it traverses AD uniformly in the unit of time ;

$\therefore AD$ represents the resultant velocity. Q.E.D.

Composition of Velocities.—The two velocities AB and AC are said to be compounded into the velocity AD.

21. *Resolution of Velocities.*—Let AB represent a velocity v. Required two velocities, of which one shall be in a given direction XY, and the other perpendicular to XY, and of which AB shall be the resultant.

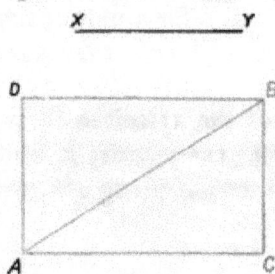

DEF. The first of these is said to be the resolved part of v in the direction XY.

Draw AC, BC, parallel, and perpendicular to XY. Complete the parallelogram CD.

FIG. 2.

The velocities represented by AC and AD will be those required.

For they are in the required directions, and, by Art. 20, they will have the velocity represented by AB for their resultant.

COR. **If** θ be the acute angle of inclination of AB to XY, $AC = v \cos \theta$; $\therefore$ *v* cos θ **is the** resolved part of v in the given direction XY.

22. Let AB, CD represent the two velocities u and v, whose directions make acute angles, θ and ϕ, with XY in opposite directions.

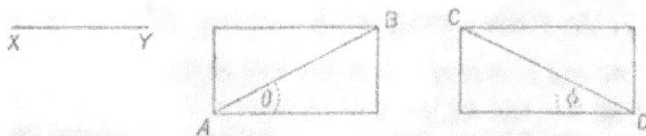

FIG. 3.

Now if amongst velocities parallel to XY those are considered positive which tend from left to right, and those negative which tend from right to left; and amongst velocities perpendicular to *XY*, those are considered positive which tend upwards from *XY*, and those negative which tend downwards; then it is easy to see that

AB is equivalent to $u \cos \theta$ parallel to *XY*,
and $u \sin \theta$ perpendicular ,, ;
CD ,, $v \cos \phi$ parallel ,, ,
and $-v \sin \phi$ perpendicular ,, ;
whilst BA ,, $-u \cos \theta$ parallel ,, ,
and $-u \sin \theta$ perpendicular ,, ;
DC ,, $-v \cos \phi$ parallel ,, ,
and $v \sin \phi$ perpendicular ,, .

23. The student will observe that there is a great similarity between the geometrical representations of forces and of velocities, and also between the compositions and resolutions of them. He must beware therefore of confusing them together. Thus he must not say that a point travels over a certain space *on account of* a particular velocity; for a velocity is not the *cause* of a point's motion, but only the rate of that motion at any instant.

EXAMPLES.—VI.

(1.) A body possesses velocities of 3 feet a second, and 9 feet a second in directions at right angles to one another. What is the resultant velocity?

(2.) If a body is moving with a velocity of 6 miles an hour in a straight line, making an angle of 30° to the due north direction, and between north and east, how fast is it moving (1) northwards, (2) eastwards.

(3.) A shot is fired from a ship with a velocity of 20 miles an hour, the gun being pointed in a direction making an angle of 45° with the ship's course, and the ship is sailing 5 miles an hour. What is the actual velocity of the shot?

(4.) The resolved parts of a velocity in two directions at right angles to one another are at any instant 2 and 3. Find the direction of its motion.

If at a subsequent instant they are 3 and 2, by how much is the direction of motion changed in the interval?

(5.) If two balls, radii r_1, r_2, be projected from points A and B, where $AB=a$, with velocities v_1, v_2, só that their directions make angles a, β with AB; show that the condition that they should just graze one another is

$$\text{sum of radii} = a \frac{v_1 \sin a \sim v_2 \sin \beta}{\sqrt{v_1{}^2 + v_2{}^2 - 2v_1 v_2 \cos (a - \beta)}}.$$

If this condition be not satisfied, find the angle which the radii, at the instant of the impact, through the point of contact make with AB.

24. *Change of Velocity.*—Let AB represent the velocity of a point at any instant, CD its velocity at any other instant.

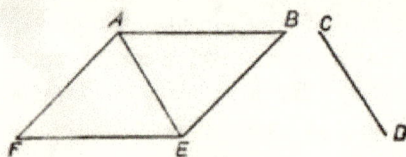

Fig. 4.

Draw AE equal and parallel to CD, and complete the parallelogram FB.

Then the velocity CD, or AE, is equivalent to the two velocities AB and AF.

Hence the velocity at the second instant is equivalent to the velocity AB, which it had at the first instant, together with an additional velocity AF.

Hence AF is said to be the *change*, or *alteration*, in the velocity of the point, during the interval between the two instants; and the velocity is said to be changed in the interval by the velocity represented in magnitude and direction by AF.

The student must remember, then, that the change in a velocity during an interval is not the difference between its magnitudes at the beginning and the end, unless the change is in the direction of the velocity at the beginning, *i.e.* unless CD is parallel to AB, and in the same direction as AB.

Thus, suppose that a point is moving in a path represented by the curve $APQB$, and from A towards B.

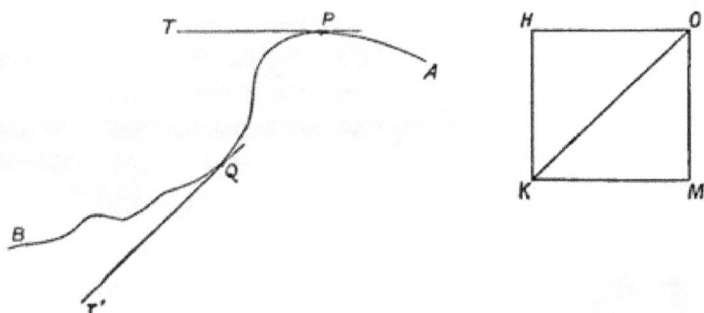

Fig. 5.

Let PT, QT' be the tangents to this curve at two points, P and Q. Then when the point is at P, it is moving for the instant in the direction PT, and when at Q, in the direction QT'.

Draw OH parallel to PT, and containing as many units of length as the velocity of the point when at P contains units of velocity.

Then OH completely represents the velocity at P.

Similarly draw OK parallel to QT' to represent the velocity at Q.

Complete the parallelogram $OHKM$.

Then, HK, or OM, represents, in magnitude and direction, the change in the point's velocity during the passage from P to Q.

26. When a point is moving in one plane with a velocity varying both in direction and magnitude, it is tedious to give both the direction and magnitude at every instant. It is easier

B

to suppose the velocity at any instant to be resolved into two parts, one parallel, and the other perpendicular, to a given direction.

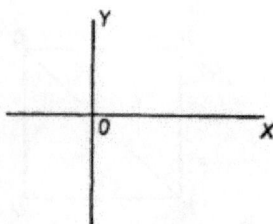

Thus let OX, OY denote two directions at right angles.

Then we should give the velocity at any instant by saying that the point had such and such a velocity (x) parallel to OX, and such and such a velocity (y) parallel to OY.

Fɪɢ. 6.

27. Suppose a point is moving with velocities, represented by AB, AC, at any two instants respectively.

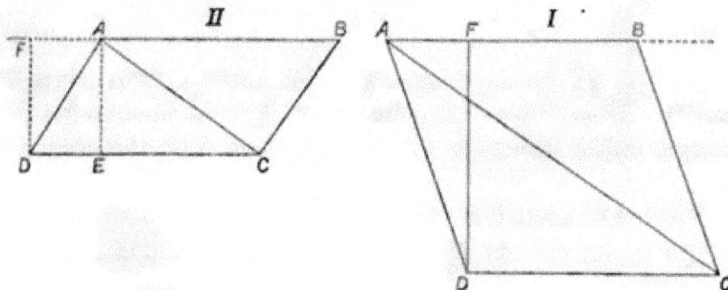

Fɪɢ. 7.

Complete the parallelogram $ABCD$; then, as before, AD represents the change in velocity during the interval between these two instants.

Draw DF perpendicular to AB, or AB produced if necessary; then the change AD is equivalent to a change represented by AF and a change represented by FD.

So that we should say that, at the second instant the point was moving with a velocity $AB\pm AF$ (+ for I, − for II) parallel to AB, and a velocity FD perpendicular to AB.

In the special case, where the whole change AD is perpendicular to AB, AF vanishes, and there is no change in the velocity parallel to AB.   Conversely, if there is no change of velocity in any particular direction, the change has taken place wholly in the perpendicular direction.

EXAMPLES.—VII.

(1.) A point is travelling at one instant with a velocity v northwards, and at another with an equal velocity eastwards.   What is the change in the interval ?

(2.) A point is travelling at one instant with a velocity of 6 miles an hour southwards, and at another with a velocity of 4 miles an hour towards the N.E.   What is the change of velocity in the interval ?

(3.) The direction of a point's motion is changed by 30°, but its rate of motion remains unaltered.   What has been the change in its velocity ?

(4.) A point is travelling at the beginning of an interval with a velocity of 5 yards a minute, and during the interval its change of velocity is one of 6 yards a minute in a direction, which makes an angle of 60° with the initial direction of motion.   What is the final motion of the point ?

II.—ACCELERATION.

28. WHEN the velocity of a point is being changed, the *rate of change of the velocity is called* the *acceleration of the velocity.*

29. The velocity of a point may be changing more rapidly at one instant than at another. ·Or when two points are in motion, and not moving each with uniform velocities, then at any instant the change in the velocity of one may be more rapid than the change in the velocity of the other. We should say, in the first case, that the acceleration of the point's velocity at one instant was greater than at the other; and in the second case, that the acceleration of the velocity of one point was greater than the acceleration of the velocity of the other. Thus we see that accelerations may differ from one another in magnitude.

30. For the phrase " acceleration of the velocity of a point," we often use the shorter one, " acceleration of a point."

31. If during any interval of time the acceleration of a point is the same at every instant, the acceleration is said to be *uniform* throughout that interval. If the acceleration at one instant is different from what it is at another, it is said to vary, or to be a *variable* acceleration.

32. If the velocity of a point always changes uniformly, *i.e.* *if its acceleration is uniform,* it is easy to see that it will acquire equal velocities in equal intervals of time. But the converse of this, viz., that, if equal velocities are acquired in equal times, the acceleration is uniform, is not necessarily true. Thus, if the acceleration of a point is uniform and of proper magnitude, it will acquire 5 units of velocity in every hour. (For instance, the point might be travelling with a velocity 8 at one moment, then an hour hence it would be travelling with a velocity 13, and at the end of a second hour it would be travelling with a velocity 18, and so on.) But if it acquires 5 units of velocity in every hour, it does not by any means follow that its acceleration is uniform throughout an hour; for at one part of that time the velocity may be changing more rapidly than at another, if only it manage to increase upon the whole exactly by 5 units of velocity in the hour. If, however, it acquired $\frac{1}{12}$ of a unit of velocity in every minute, we should feel more confident that the velocity was changing uniformly, and still more so if it acquired $\frac{1}{720}$ of a unit of velocity in every second; and if, on dividing the time into equal intervals, as small as we pleased, we found that the point acquired an equal velocity in each interval, we should conclude that the acceleration was uniform throughout the whole time. We thus arrive at the following *test* for uniformity of acceleration :—

An acceleration of a velocity of a point is uniform when equal velocities are acquired in equal intervals of time, *however small*.

33. The student must notice the difference in meaning between the phrases, " the velocity possessed at any instant," and " the velocity acquired during any interval."

Thus, in the instance given in Art. **32**, the point possesses at first a velocity 8, at the end of the hour it possesses a velocity 13, and at the end of the second hour it possesses a velocity 18; and it has acquired during each hour a velocity 5.

34. We can indicate the magnitude of an acceleration by the *phrase* expressing the number of units of velocity which a point **would** acquire in **some stated** interval of time, during which the velocity changed uniformly with the acceleration indicated.

Thus, we talk of **an** acceleration of " 7 units of velocity per **hour**," meaning that a velocity would change by 7 units in any **hour**, during which it changed uniformly with the acceleration indicated.

Now, if a foot and a second were the units of space **and** time, the phrase " 7 units of velocity" would be equivalent to the phrase a velocity of " 7 feet per second" (Art. 9). Hence we could indicate the above acceleration by **the** phrase " 7 feet per second per hour."

Again, in one second $\frac{7}{3600}$ of a **unit of** velocity would be acquired with **the** above acceleration. Hence it might also be indicated by the phrases, " $\frac{7}{3600}$ of a unit of velocity per second," and " $\frac{7}{3600}$ of a foot per second per second."

Ex. 1. **A point** possesses an acceleration of 20 units of velocity per 75'. What velocity will it acquire in an hour ?

The point **acquires 20** units of velocity in 75' ;

$$\therefore \text{ it} \qquad \text{,,} \qquad \frac{20}{75} \qquad \text{,,} \qquad \text{,,} \qquad 1' ;$$

$$\therefore \text{ ,,} \qquad \text{,,} \qquad \frac{20}{75} \times 60 \qquad \text{,,} \qquad \text{,,} \qquad 60' ;$$

$$\textit{i.e.} \text{ ,,} \qquad \text{,,} \qquad 16 \qquad \text{,,} \qquad \text{,,} \qquad 1 \text{ hour.}$$

Ex. 2. A point has the acceleration 7 feet per second per minute. What velocity will it acquire in an hour ?

In one minute it acquires the velocity of 7 feet per second, and in 60' it acquires a velocity 60 times as great.

Now, the velocity of 7 feet per second is that with which a point would **go 7 feet in a second ; and** with the velocity 60 times **as** great, a point would go 60 times **as far in a** second, *i.e.* 420 feet in a **second.** Hence the given point acquires in an hour the velocity of 420 feet per second.

(1.) A point is travelling with an acceleration of 20 units of velocity a minute. What velocity will it acquire in an hour?

(2.) What velocity will a point acquire in half an hour whose acceleration is 15 feet per second per second?

(3.) A point is travelling with the acceleration of 12 feet per second per hour. What will be the change in its velocity in a minute?

(4.) A point's acceleration is the velocity 3 per second. What velocity per hour would represent this acceleration?

(5.) A train's acceleration is 5 feet per second per second. How long will it take to acquire a velocity of 100 yards per minute?

(6.) A train acquires the velocity of 30 feet per second in an hour. If its motion is uniformly accelerated, what velocity will it acquire in a minute?

36. By one acceleration being double another we understand that, if two velocities increase uniformly for any the same interval of time, one with one acceleration and the other with the other, then at the end of the interval the first velocity would have increased by twice as much as the other.

Thus, suppose the velocities of two points originally were 3 and 4, and, at the end of an hour, were 13 and 9; then, if they increased uniformly, the acceleration of the first point would be said to be double that of the second at any instant.

Generally, if the acceleration of one velocity is a times that of another, then during any given interval the change in the first velocity would be a times that in the second; and we should say that the acceleration of the first *contained* the acceleration of the second a times.

37. *Ex.* 1. Show that the acceleration of 360 feet per second per hour is double that of 1 yard per second per minute.

With the first a point would acquire in one minute a velocity of 6 feet per second, *i.e.* 2 yards per second, and this is double the velocity of 1 yard per second.

Hence in one minute a point would acquire with the first acceleration **double the** velocity that it would **with** the second.

Hence the first acceleration is double the **second.**

Ex. 2. Again **the acceleration** of 30 **yards per minute per minute is** ¼ of the first of the above accelerations.

For with it a point would acquire in one minute a velocity of 30 yards per **minute,** which is ½ **yard per** second, which is ¼ of the velocity 2 yards per second.

Hence with this third acceleration a point would in one minute acquire a velocity ¼ **of** what it would acquire in the same time with the first acceleration, *i.e.* the third acceleration **is one quarter** of the first.

EXAMPLES.—IX.

(1.) A point **acquires 5** units of velocity per second, and another, 20 units per **minute.** Compare their accelerations, supposing them to be uniform.

(2.) The acceleration **of** 300 yards per minute per minute is a quarter **of** the acceleration of one foot per second per second.

(3.) The acceleration x feet per second per minute **is** double that of 20 yards per minute per second. Find x.

(4.) Show that the acceleration of a foot per second per minute is **equal to that** of a foot per minute per second.

(5.) **Find** the ratio between two accelerations, one being 10 feet per **second per** minute, and the **other 300** yards per minute per minute.

(6.) **How many** units of velocity **an** hour must an acceleration be, in order **to** be one-third of one of 7 **units of** velocity a minute?

(7.) **How many** times does **the** acceleration of 6 units of velocity **per** minute **contain** that of 2 **units** of velocity per second?

(8.) How many times does the acceleration of 5 feet per second per second contain that **of** 720 feet per minute per minute?

38. In *Algebra* an acceleration must **be** represented by a number, *i.e.* by its measure (Alg. **Pt. I. Art.** 33), viz., the number of times it contains some known standard.

For the phrase "the acceleration whose measure is a," we often write "the acceleration a;" and we must then remember that this only means that a is the *number* of times which the acceleration indicated contains (Art. 36) the standard acceleration.

The particular acceleration we fix upon as our standard is that with which the velocity of a point will increase by a unit of velocity in a unit of time.

39. PROP. *The measure of an acceleration is equal to the measure of the velocity acquired by a point in a unit of time, when the velocity of the point is changing uniformly with that acceleration.*

Let a denote the measure of the acceleration, *i.e.* let it contain the standard a times. Therefore, Art. 36, a velocity, changing with this acceleration, would increase by a times as great a velocity in one unit of time as if it were changing with the standard.

Now with the standard the velocity would increase by one unit of velocity in a unit of time, and therefore with the acceleration a it will increase by a units of velocity in a unit of time, or in other words, a is the measure of the velocity the point acquires in a unit of time. Q.E.D.

COR. Suppose a velocity of a point changes *uniformly* for a time t with an acceleration a. By the Prop. the point acquires a units of velocity in each unit of time, hence in t units of time it acquires a velocity at.

So that if v' denote the velocity *acquired* in time t, we have

$$v' = at, \text{ or } \frac{v'}{t} = a.$$

Conversely, if the fraction $\frac{v'}{t}$ is always the same for all values of t (*i.e.* if $v' \propto t$), we conclude that the acceleration is constant, or uniform. This includes the test of Art. 32.

Also, if u be its velocity at the beginning of the time t, its velocity, v, at the end $= u + v'$, or $v = u + at$.

40. *Ex.* 1. If the unit of time be a minute and a foot the unit of space, what is the measure of the acceleration of 40 miles per hour per hour?

A point having this acceleration acquires in one hour

a velocity of   40   miles an hour.

i.e.   ,, $40 \times 1760 \times 3$ feet   ,,

i.e.   ,, $\dfrac{40 \times 1760 \times 3}{60}$,,   a minute,

i.e.   ,,   3520   ,,   ,,

Now, a foot and a minute being units of space and time, the velocity of 3520 feet a minute contains 3520 units of velocity.

Hence the point acquires 3520 units of velocity per hour.

i.e. $\dfrac{3520}{60}$,,   ,,   minute,

i.e. $\dfrac{176}{3}$,,   ,,   ,,

Hence the measure of the acceleration is $\dfrac{176}{3}$.

Ex. 2. What velocity will a point having an acceleration 3 acquire in 4 units of time?

With the given acceleration

     it acquires 3 units of velocity in 1 unit of time;

∴   ,,   12 ,,   ,,   4 units ,,

EXAMPLES.—X.

(1.) What is the measure of the acceleration 70 units of velocity a second, when a second is the unit of time?

(2.) What is the measure of the acceleration 3 feet per second per second, when a foot and a second are the units of space and time?

(3.) What velocity is acquired in a minute by a point whose acceleration is 3, a minute being the unit of time?

(4.) What velocity is acquired in an hour by a point whose acceleration is 11 feet per second per second, a foot and a second being units of space and time?

(5.) An acceleration of one unit of velocity per 7 seconds is the standard acceleration. What is the measure of the acceleration 15 units of velocity per 7 seconds ?

(6.) What will be the measure of the acceleration of 7 feet per second per second, when a yard per minute per minute is the standard ?

(7.) What velocity must a point acquire in an hour in order that the measure of its acceleration may be $\frac{25}{9}$, when a foot and a second are the units of space and time ?

(8.) How many minutes will a body take to acquire a velocity 100 with an acceleration 5 ; the standard being the acceleration of 25 units of velocity per 3' ?

(9.) If 5 yards and 3' are the units of space and time, what velocity will a point acquire in half an hour with the acceleration 7 ?

(10.) Two points move, one with the acceleration of 5 yards per second per second, and the other with the acceleration 7 feet per minute per minute. By how much will their velocities differ (1) at the end of 5', when they start simultaneously ; (2) at the end of 7' from the starting of the last, when the first starts 3' before the second ?

(11.) If half a minute be taken as the unit of time and a yard as the unit of space, find the numerical value of the acceleration 32 feet per second per second.

(12.) If the acceleration of 30 feet per second per second be repre-sented by 5, what will be the measure of the acceleration 7 yards per minute per minute ?

41. If the velocity of a point is decreasing, or being retarded, all that we have said is true, except that we should read " decrease " for " increase," " lose " for " acquire." It is better, however, in order that the propositions may suit both cases, to consider that when the velocity in a positive direction is decreasing, it has an acceleration in an opposite direction, which is represented by a negative symbol.

And more generally we adopt the following convention :—

Let velocities be considered positive when points are moving from left to right, and negative, from right to left. Then, if a positive velocity is being increased, it is said to have a posi-tive acceleration, if it is being decreased to have a negative

acceleration. If a negative velocity is being numerically increased (*i.e.* algebraically decreased), it is said to have a negative acceleration, and if numerically decreased (*i.e.* algebraically
increased), to have a positive acceleration.

42. Thus, if a point have at any instant a velocity 10 from left
to right, and at the end of a second a velocity 4 from left to
right, and at the end of another second a velocity 2 from right
to left, we should say that it had at these three instants velocities 10, 4, and −2 respectively; and if further the change took
place uniformly, we should say it had an acceleration −6
throughout the two seconds.

Again, if it had been moving at first with a velocity 12 from
left to right, and at the end of a second with a velocity 5 from
left to right, and at the end of another second with a velocity
2 from right to left, and if these changes took place uniformly,
we should say that it had, at the three instants, velocities
−12, −5, and 2 respectively, and that its acceleration
throughout the two seconds was 7.

In both cases we have considered a second to be the unit of
time.

43. With the understanding that our symbols for velocities,
and accelerations, may be positive, or negative, the student will
find the following statement to be true :—

If a point have a uniform acceleration a for a time t, and
if u be its velocity at the beginning of the time, and v its
velocity at the end, then

$$v = u + at.$$

44. The two characteristics of an acceleration are its magnitude and its direction. An acceleration cannot be said to be
completely determined until both these characteristics are
determined.

45. Accelerations can be represented by straight lines.

For a straight line can be drawn

1° in any direction, and thus can represent any acceleration in direction;

2° so as to contain as many units of length as the acceleration contains units of acceleration, and thus can represent the acceleration in magnitude.

46. Since the number of units in any acceleration is the same (Art. 39) as the number of units of velocity, which would be acquired in a unit of time with that acceleration; hence the acceleration of a point's motion at any instant, and the velocity, which the point would acquire with that acceleration in a unit of time, are represented by the same straight line.

EXAMPLES.—XI.

(1.) A point has at one instant a velocity 5. What velocity has it 3 minutes afterwards, supposing it to move with an acceleration of 2 feet per second per second, in the direction of its initial motion, a foot and a second being units of space and time?

What would be its velocity at the end of the 3 minutes, if the acceleration were in the direction opposite to that of its initial motion?

(2.) An acceleration 5 is represented by a line 10 inches long. What length of line will represent an acceleration 9?

(3.) If the acceleration of 4 miles per minute per minute towards the north be represented by 5, how would you represent the acceleration of 2 feet per second per second towards the south?

(4.) Given that a certain line containing 3 inches represents the acceleration 8 miles per hour per hour towards the east; how would you represent the acceleration of 100 yards per minute per minute towards the north-east?

47. We have defined our unit of acceleration with reference to the units of velocity and time. Hence, if these be changed, the acceleration taken as our standard must generally be different, and therefore the measure of each particular acceleration will be changed.

Further, the unit of velocity depends on the units of space and time. Hence, if these be changed, we must generally take for our standard a different acceleration, and then the measure of each acceleration will be changed.

48. PROP. *Given the measure of any acceleration with certain (old) units of space and time, to find the measure of the same acceleration with any other (new) units of space and time.*

Let a be the measure of the acceleration with the old units.

Let a and b denote the number of times, respectively, which the new units of space and time contain the old.

With the acceleration we are considering a point will acquire

a old units of velocity in one old unit of time ;

i.e., Art. 16, $\dfrac{ba}{a}$ new units of velocity in one old unit of time ;

and $\therefore$ $b \cdot \dfrac{ba}{a}$ new units of velocity in one new unit of time ;

but, by Prop. Art. 39, the measure of the velocity acquired in a unit of time is the measure of the acceleration ;

$\therefore$ $\dfrac{b^2 a}{a}$ is the new measure (a') of the acceleration required.

Hence, if the values of any three of the symbols, a, a', a, b, are known, the value of the fourth is found from the equation

$$a' = \frac{ab^2}{a}.$$

49. *Ex.* 1. What is the measure (a') of the acceleration 7 feet per minute per minute, when a yard and a second are taken as the units of space and time ?

If a foot and a minute were the old units of space and time, the measure of the velocity acquired in a minute would be 7, and therefore the measure of the acceleration would be 7 (Art. 39).

Also the **new units** of space and time given in the question are respectively 3 times and $\frac{1}{60}$ of these old units;

$\therefore$, putting $a=7$, $b=\dfrac{1}{60}$, $a=3$, we have

$$a' = \frac{\overline{\dfrac{1}{60}}\Big|^{2} \cdot 7}{3} = \frac{7}{10800}.$$

Ex. 2. If the acceleration 10 feet per second per second is represented by 12, and a minute is the unit of time, find the unit of space.

If a foot and a second were the units of space and time, the measure of the above acceleration would be 10.

Let a feet be the new unit of space.

Then if a feet and one minute are the units, the measure of the acceleration is 12; also $1' = 60''$.

Hence we can put $a=10$, $a'=12$, $b=60$;

$$\therefore 12 = \frac{\overline{60}\big|^{2} \cdot 10}{a} = \frac{36000}{a};$$

$$\therefore a = 3000.$$

Hence a length of 3000 feet is the unit of length.

EXAMPLES.—XII.

(1.) What is the measure of the acceleration 32 feet per second per second, when $3''$ and 3 yards are the units of time and space?

(2.) If a minute be taken as the unit of time and a velocity of 60 miles an hour as the unit of velocity, what will be the measure of the acceleration, whose measure is 32, when a foot and a second are taken as the units of space and time?

(3.) If 3 be the numerical value of that acceleration with which in 3 seconds a velocity of 3 feet per second is acquired, what has been taken as the unit of space, if 5 seconds be the unit of time?

(4.) If an acceleration be represented by the same number when the units of time are $1''$ and $2''$, find the ratio between the corresponding units of length.

(5.) If g be the measure of an acceleration when m seconds and n feet are the units of time and space, show that the measure of the acceleration, when n seconds and m feet are the units, is $g\dfrac{n^3}{m^3}$.

(6.) The measures of an acceleration and a velocity when referred to $(a+b)$ ft., $(m+n)''$ and $(a-b)$ ft., $(m-n)''$ respectively are in inverse ratio of their measures when referred to $(a-b)$ ft., $(m-n)''$ and $(a+b)$ft., $(m+n)''$; their measures when referred to a ft., m'' and b ft., n'' are as $ma : nb$. Show that

$$\frac{n}{m} = \sqrt{1 - \frac{b^4}{a^4}}.$$

(7.) If the unit of velocity be the velocity with which a point passes over a feet in t seconds, and the unit of acceleration that of a point which acquires in t seconds a velocity of b feet in t seconds, find the units of space and time.

(8.) Show that the unit of space varies, directly as the unit of space, and in the inverse duplicate ratio of the unit of time.

(9.) A point, having a certain acceleration, in $8'$ acquires a velocity represented by 5 feet per 3 seconds, and, when the unit of length is $\frac{1}{8}$ of what it was before and 4 seconds the unit of time, the measure of the acceleration is 20 ; find the number of feet in the unit of length in the first case.

(10.) If the acceleration of $32 \cdot 2$ feet per second per second be taken as the unit of acceleration, and the velocity of $32 \cdot 2$ feet in $16 \cdot 1$ seconds as the unit of velocity, what are the units of space and time respectively?

50. We have seen, in Art. **18**, how a point may be considered as having, at any instant, two or more velocities in different directions.

We may now go further and suppose that the velocity in any one, or more, of these directions is being changed. Then the rate at which it is being changed is called the acceleration of the velocity of the point in that direction ; or more shortly, though inaccurately, the acceleration of the point in that direction.

For example, it might be convenient to regard the velocity of a point, at any instant, as made up of two components, one in direction XX' and one in direction YY'. Then the rates,

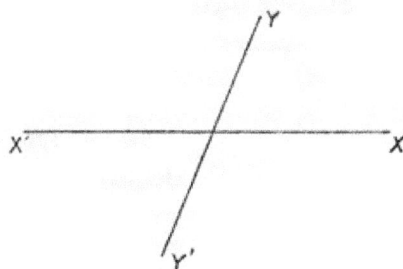

FIG. 8.

at which these velocities are changing at the instant under consideration, would be called the accelerations of the point in these directions at that instant.

Further, it might happen that at any instant the actual value of the velocity in the direction XX' might be *nil ;* so that the point would be moving in direction YY', but might possess accelerations in the directions both of YY' and of XX'.

The student will thus see how a point can be considered as having accelerations in different directions at once, and also that it may at the same instant have a direction of actual motion different from any of these directions.

51. In Art. **50,** suppose at any instant that u and v were the velocities in the directions XX', YY', and that for a time t the point possessed uniform accelerations, a, β, in these directions. Then at the end of the time t its velocities would be

$$u + at \text{ in direction } XX', \text{ and}$$
$$v + \beta t \quad \text{,,} \qquad YY'.$$

So that the actual velocity at the beginning of the time would be found by compounding u and v according to the Parallelogram Law in Art. **20**; and, at the end of the time, by compounding $u + at$, $v + \beta t$.

C

52. DEF. When a point possesses at any instant two or more accelerations, the actual acceleration is called their *resultant,* and they are called the *components* of this resultant.

53. PROP. The Parallelogram of Accelerations.—*If two straight lines AB, AC, represent two accelerations possessed by a point at any instant, and the parallelogram ABDC be completed, then AD will represent the resultant acceleration.*

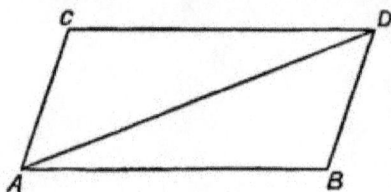

Since *AB, AC* represent the accelerations, they represent also (Art. **46**) the two velocities which would be acquired in a unit of time with those accelerations. That is to say, at the end of this unit of time any point would, with these accelerations, have acquired two additional velocities represented by *AB* and *AC.* The resultant of these velocities, Art. **20**, is represented by *AD*; ∴ *AD* represents the resultant change of velocity acquired in a unit of time, and ∴, Art. **46**, it represents the resultant acceleration.　　　　　Q.E.D.

FIG. 9.

Composition of Accelerations.—The two accelerations *AB* and *AC* are said to be compounded into the acceleration *AD.*

54. *Resolution of Accelerations.*—Let *AB* represent any acceleration, then the *resolved part* of *AB* in any direction, *XY,* is the name given to the acceleration in the direction *XY,* such that it and a certain other, perpendicular to *XY,* have *AB* for the resultant.

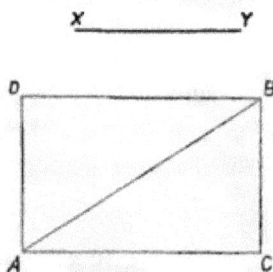

Describe the rectangle *ACBD* having one side *AC* parallel to *XY,* and *AB* for a diagonal. Then *AC* and *AD* have *AB* for their resultant, and therefore *AC* represents the resolved part of *AB* in the direction *XY.*

FIG. 10.

If a represents AB, and θ the acute angle it makes with XY, then $a \cos\theta$ is the measure of this resolved part.

The sign to be prefixed to $a \cos\theta$ must be determined as in Art. 22.

EXAMPLES.—XIII.

(1.) A point's motion is changing with the accelerations 3 and 4 in two directions at right angles. Determine the resultant acceleration.

(2.) A point has an acceleration 10 towards the east, and another 12 towards the north. At noon it is moving with a velocity 100 towards the east. How is it moving at 12.15 P.M., a foot and a second being the units of space and time?

(3.) A point starts with a velocity 60, and always has an acceleration 3 in a direction making an angle 30 with its initial direction of motion. How is it moving at the end of 10 units of time after the instant of starting?

(4.) At one instant a point is moving with a velocity of 5 feet per second, and (having a constant acceleration all the time) 10" afterwards it has an equal velocity in the direction perpendicular to the former. What is the acceleration?

55. SUPPOSE that during any interval of time t, the velocity of a point is uniformly changing in magnitude with an acceleration a, whilst the point traverses a space s in the direction of this velocity, and that at the beginning and end of the interval the measures of the velocity are u and v.

Then we have already found, in Art. **39**, Cor., the relation

$$v = u + at.$$

We proceed to find the relations into which s enters.

56. Divide the interval t into n equal intervals, of each of which the length is τ, so that $t = n\tau$.

The velocities at the beginnings of these intervals will be

$$u,\ u + a\tau,\ u + a.2\tau,\ \ldots,\ u + a.\overline{n-1}\tau\ ;$$

and the velocities at their ends will be

$$u + a\tau,\ u + a.2\tau,\ \ldots,\ u + a.\overline{n-1}\tau,\ u + a.n\tau.$$

I. Suppose the point to move during each interval with the velocity it has at the beginning of that interval, then the whole space passed over would be

$$u\tau + (u + a.\tau)\tau + (u + a.2\tau)\tau + \text{ etc. } + (u + a.\overline{n-1}.\tau)\tau$$

$$= u.n\tau + a\tau^2(1 + 2 + \ldots + \overline{n-1})$$

$$= un\tau + \frac{a\tau^2}{2}n(n-1)$$

$$= ut + \frac{at^2}{2}\left(1 - \frac{1}{n}\right).$$

II. Suppose the point to move during each interval with the velocity it has at the end of that interval, then the whole space passed over would be

$$(u+a\tau)\tau+(u+a.2\tau)\tau+ \text{ etc. } +(u+a.n\tau)\tau$$
$$=un\tau+a\tau^2(1+2+ \ldots +n)$$
$$=u.n.\tau+\frac{a\tau^2}{2}n(n+1)$$
$$=ut+\frac{at^2}{2}\left(1+\frac{1}{n}\right).$$

Now the space (s) actually passed over must lie between these two expressions, whatever n may be, and since when n is endlessly increased they both become $ut+\dfrac{at^2}{2}$, we have

$$s=ut+\frac{at^2}{2}.$$

57. We have $v=u+at$, or $at=v-u$,

$$\text{and } s=\left(u+\frac{at}{2}\right)t;$$
$$\therefore s=\left(u+\frac{v-u}{2}\right)\frac{v-u}{a}$$
$$=\frac{v+u}{2}\cdot\frac{v-u}{a}$$
$$=\frac{v^2-u^2}{2a};$$
$$\therefore 2as=v^2-u^2, \text{ or } v^2=u^2+2as.$$

58. These three,

$$v=u+at, \qquad \qquad \text{(1)},$$
$$s=ut+\tfrac{1}{2}at^2, \qquad \text{(2)},$$
$$v^2=u^2+2as, \qquad \text{(3)},$$

are the fundamental formulæ for uniformly accelerated motion. We will give some examples of their use.

59. A case of uniformly accelerated motion, which very frequently occurs in nature, is that of a body moving vertically upwards or downwards.

When a particle is falling vertically downwards towards the earth, it is found by experiments, some of which will be subsequently explained, that the particle has a uniform acceleration, such that during every second of its motion it acquires an additional velocity of 32·2 feet per second.

Thus we say that the acceleration of a falling body is

32·2 feet per second per second downwards;

and that, a foot and a second being the units of space and time, the measure of the acceleration of a falling body is 32·2 downwards. This number we generally represent by g.

A particle thrown vertically upwards has the same acceleration. So that, if we represent the velocity with which it starts when thrown upwards by a positive symbol, we must represent its acceleration by $-g$, a foot and a second being units.

60. *Ex.* 1. A point is thrown upwards with the velocity of 100 feet per second; find,

    (1.) Its velocity at the end of $2''$;
    (2.) The time of rising to its greatest height;
    (3.) Its velocity at the end of $4''$;
    (4.) When it will be at a distance of 100 feet from the point of starting;
    (5.) Its velocity when at a distance of 150 feet from the point of starting.

Here we put $u=100$, $a=-32\cdot2$, or $-g$.

(1.) The velocity (v) after $2''$ is obtained from the formula $v=u+at$, by putting $t=2$;

$$\therefore\ v=100-(32\cdot2)2=35\cdot6;$$

i.e. at the end of two seconds it has the velocity of 35·6 feet per second upwards.

(2.) Let t be the number of seconds which elapse before it comes to its greatest height. When it is at this height, it is at rest; ∴ in the formula $v = u + at$, by putting $v = 0$, we have

$$0 = 100 - (32{\cdot}2)t ; \therefore t = \frac{100}{32{\cdot}2} = 3\tfrac{17}{161} ;$$

∴ it takes $3''\tfrac{17}{161}$ to reach its greatest height. It then begins to descend.

(3.) Put $t = 4$; its velocity (v) at the end of $4''$ is given by

$$v = 100 - (32{\cdot}2)4 = -28{\cdot}8 ;$$

i.e. it then has the velocity 28·8 feet per second downwards.

(4.) Let t be the number of seconds which elapse before it is at a distance of 100 feet from point of starting.

Then, in the formula $s = ut + \tfrac{1}{2}at^2$, putting $s = 100$, we have

$$100 = 100t - \tfrac{1}{2}(32{\cdot}2)t^2,$$
$$\text{or } (16{\cdot}1)t^2 - 100t = -100.$$

Solving this quadratic in t, we have

$$t = \frac{50 \pm 10\sqrt{8{\cdot}9}}{16{\cdot}1} .$$

Giving the square root the approximate value 3, for the sake of example, we have $t = 4\tfrac{156}{161}$, or $1\tfrac{39}{161}$.

Thus it will be at a distance of 100 feet from the point of starting after $1''\tfrac{39}{161}$, and again after $4''\tfrac{156}{161}$, from the time of starting, approximately.

(5.) Let v be its velocity when at a distance 150 feet from point of starting.

Then, in the formula $v^2 = u^2 + 2as$, putting $s = 150$, we have

$$v^2 = (100)^2 - 2(32{\cdot}2)150 = 340 ;$$
$$\therefore v = \pm \sqrt{340}.$$

Hence it is twice at a height of 150, once going upwards, and once coming downwards, and at both times the magnitude of its velocity is the same, viz. $\sqrt{340}$ feet per second.

Ex. 2. A point starts from rest, and has the acceleration 40 feet per second per second, find the distance it traverses in the fourth half-second of its motion.

By the phrase "starts from rest" is meant that its initial velocity (u) is zero.

Take a foot and a second as units of space and time.

Let x be the number of **feet** traversed in $1\frac{1}{2}'$ from rest.

$\qquad x' \qquad\qquad\qquad,, \qquad\qquad\qquad,, \qquad\qquad 2'' \qquad\quad ,,$

Then $x'-x$ represents the distance required.

In the formula $s=ut+\frac{1}{2}at^2$, putting,

(1.) $s=x$, $u=0$, $a=40$, $t=\dfrac{3}{2}$, we have

$$x=\frac{1}{2}\cdot 40\cdot\frac{9}{4}=45\;;$$

(2.) $s=x'$, $u=0$, $a=40$, $t=2$, we have

$$x'=\frac{1}{2}\cdot 40\cdot 4=80\;;$$

$$\therefore\; x'-x=35\;;$$

i.e. the point traverses 35 feet in the given interval.

EXAMPLES.—XIV.

[A foot and a second are taken as the units of space and time unless it is otherwise stated.]

(1.) A particle drops vertically from rest. What will be its velocity (1) at the end of $10''$; (2) when it has traversed 60 feet? How far will it go (3) in the first $10''$ of its motion; (4) in the second $10''$? (5) How far must it go before it has a velocity 100?

(2.) A point is thrown downwards with a velocity $64\cdot4\;(=2g)$. How far must it go before it has acquired a velocity $4g$?

(3.) A body starts with a velocity 5, and has a constant acceleration 10 in the direction of its motion. How far will it go in $10''$? How long will it take to go 10 ft.?

(4.) A body starts with a velocity 15, and has a constant acceleration 5 in the opposite direction. When and where will it come to rest?

(5.) A body is **thrown** upwards with a velocity 20. When will it have a velocity whose magnitude is represented by 30?

(6.) A body starts with a certain velocity and moves with a constant acceleration 10 in the direction of its initial motion. In $5''$ it has traversed 450 feet. What is its initial velocity?

(7.) A point, starting with a velocity of 40 feet per second, traverses 300 feet in 5" with a constant acceleration in the direction of initial motion. What is the magnitude of this acceleration?

(8.) A point moves with uniform retardation over 100 feet in 12", starting with a velocity 10. When and where will it come to rest?

(9.) A point has described 30 feet from rest in 3". Find the acceleration (supposed uniform) and the velocity acquired.

(10.) A small body thrown upwards passes the point 20 feet from the point of starting with a velocity 55 feet. How much farther will it go, and what was the velocity with which it was projected?

(11.) Two bodies fall from heights of 20 and 30 feet, and reach the ground simultaneously. What was the interval between their starting?

(12.) A heavy particle is dropped from a given height h, and at the same instant another particle is thrown vertically upwards so that they meet half way. Find the velocity of projection of the latter particle.

(13.) A stone is dropped into a well, and after 3 seconds the sound of the splash is heard. Find the depth of the well, supposing that the velocity of sound 1000 feet per second, and that the stone will fall 16·1 feet in the first second.

(14.) If S_r be the space described by a body, moving with a uniform acceleration a, in the rth unit of time, show that $\dfrac{2S_r}{a}$ must be an odd integer.

(15.) During any uniformly accelerated motion, the space described in any interval of time is the same as might be described by a body moving uniformly with the mean of the extreme velocities.

(16.) The path of a body uniformly accelerated is divided into a number of equal spaces. Show that, if the times of describing these spaces be in $A.P.$, the sums of the greatest and least velocities for each such space respectively are in $H.P.$

(17.) A body moves from rest with an acceleration, which remains constant during certain successive equal intervals of time, but is changed at the expiration of each such interval, so that the space described in the nth interval is always $\dfrac{2^{n+1}-3}{2^{n-1}}$ times the space described in the first of them. If the velocity acquired at the end of the first interval be v, show that after a long lapse of time the velocity approaches a uniform velocity $2v$.

(18.) One particle describes the diameter AB of a circle with uniform velocity, and another the semi-circumference AB with uniform tangential acceleration, they start together from A and arrive together at B; show that the velocities at B are as $1 : \pi$.

(19.) A particle falls freely in the nth of a second through a space a; what must be the unit of length?

(20.) If the unit of velocity be that which describes m units of space in the unit of time, and the unit of acceleration be that which produces n units of velocity in the unit of time, how must the equation $s = vt + \frac{1}{2}at^2$ be modified?

(21.) A body moves from rest with uniform acceleration over 15 yards in 3 seconds, find the space described in the last second.

(22.) A particle moves with a constant acceleration a. If u be the arithmetic mean of the first and last velocities in passing over any portion h of the path, and v the velocity gained, show that $uv = ah$.

(23.) If a body describes 36 feet whilst its velocity increases uniformly from 8 to 10 feet per second, how much farther will it be carried before it attains a velocity of 12 feet per second? Required also the magnitude of the acceleration.

(24.) A point, moving with a uniform acceleration, describes 20 feet in the half second which elapses after the first second of its motion. Compare its acceleration with that of a falling particle, and give its numerical measure, taking a minute as the unit of time, and a mile as that of space.

(25.) If a body be projected upwards with a velocity ng (where g is the measure of the acceleration of gravity), when will its height be ng, and what will then be its velocity?

Show that n must not be less than 2.

(26.) A body describes 100 feet from rest, and acquires a velocity of 40, with a uniform acceleration 8, in 5 seconds, what have been taken as the units of space and time?

(27.) A point A starts from a given point with uniform acceleration, and afterwards another B from the same point with a greater acceleration; show that the difference of the final velocities when B overtakes A is to the less of the two, as A's velocity when B started is to the velocity acquired by A in the remainder of the motion.

(28.) A point moves with uniform acceleration, and describes 12 feet in the second half-second of its motion, and, at the end of that half-second, is moving with the unit of velocity; if a yard be the unit of space, determine that of time and the numerical value of the acceleration.

(29.) A body falls through a feet and acquires a velocity b, with a uniform acceleration f, in t seconds, what are the units of time and length ?

(30.) A body is observed to describe in successive intervals of $2''$ each the spaces 24 and 64 feet in the same straight line ; show that the acceleration may be constant, and find its measure.

(31.) A body starts from rest under a uniform acceleration, but at the commencement of each successive second the acceleration is decreased in a geometrical proportion $(r = \frac{1}{2})$; show that the space described in n seconds $= \left(2n - 3 + \dfrac{3}{2^n} \right) 2s$, where s is the space described in the first second.

(32.) If v_n be the velocity, at the end of n seconds, of a body which has an initial velocity, and a uniform acceleration, and S_n be the space described during n seconds, show that

$$\frac{S_{n+1}}{n+1} - \frac{2S_n}{n} + \frac{S_{n-1}}{n-1} = v_{n+1} - 2v_n + v_{n-1}.$$

61. Suppose that a point moves so as always to remain in one plane, and that its motion has a constant acceleration.

Let OX, OY be two fixed straight lines in this plane. Then, since the acceleration is constant, it can always be resolved

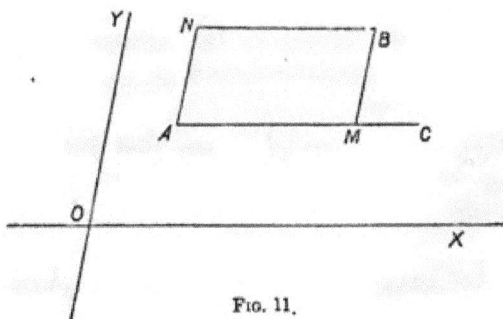

Fig. 11.

into the same two components parallel to OX and OY. Let a, β be their respective values. Let A be the position of the point at any instant, and B after an interval t. Let its velocity when at A be equivalent to the two components u, v, parallel to OX and OY.

Draw AC parallel to OX.

Then we can make the point execute its motion by supposing it to move along AC with an acceleration a, starting from A with a velocity u, whilst AC remains parallel to itself, with A moving parallel to OY with an acceleration β, starting from its initial position with a velocity v. For then the point will have its proper accelerations in these directions, and its proper initial velocities.

· Draw BM parallel to OY, meeting the initial position of AC in M, and complete the parallelogram $AMBN$.

Then BM, or AN, is said to be the distance of B from the initial position of A measured *parallel to* OY, and AM, or BN, its distance *measured parallel to* OX.

Now when AC has arrived in the position NB, *i.e.* when A has travelled over AN from its original position to N, the point is at B, *i.e.* has travelled along AC over a distance AM.

Hence, $AM = ut + \frac{1}{2}at^2$,
$$AN = vt + \frac{1}{2}\beta t^2.$$

Also its velocity at B will be equivalent to two components, one, parallel to OX, $= u + at$,
and the other, parallel to OY, $= v + \beta t$.

Hence we can find, separately, the motions of the point at any instant, and the spaces traversed by it in any time, parallel to two given directions.

The student must be careful to note that the point does not travel along the straight line AB, unles $a = 0$, and $\beta = 0$, or $u = 0$, and $v = 0$.

62. The following are particular cases to which we shall recur :—

(1.) Let $v = 0$, and $a = 0$. Then $AM = ut$,
$$AN = \frac{1}{2}\beta t^2.$$

Here AM is the direction of motion of the point when at A, and OY is the constant direction of the acceleration of the motion of the point throughout the whole interval.

Thus the point starts from A with a velocity of which the direction is not coincident with that of the constant acceleration.

(2.) Let XOY be a right angle, and $v=0$.

Then $AM = ut + \frac{1}{2}at^2$

$AN = \frac{1}{2}\beta t^2$.

Here AM is the direction of motion of the point when at A; and the point can be regarded as having two co-existing accelerations, one in the direction of motion at A, and the other perpendicular to it.

IV.—ANGULAR VELOCITY AND NORMAL ACCELERATION.

63. *Angular Velocity.*—Let A be a fixed point, AX a fixed straight line, and let P be a moving point.

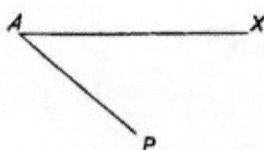

Then, if P moves so that the angle PAX is being changed, P is said to have an *angular motion* with respect to, or about, A; and the rate at which the angle PAX is being changed is called the *angular* velocity of P with respect to A.

FIG. 12.

The only way in which P can move, so as not to have this angular motion, is along AP.

64. The unit of angular velocity is that angular velocity with which a moving point will pass through a unit of angle in a unit of time, about a fixed point.

65. As in Art. **9** it can be shown that, if ω be the measure of any particular angular velocity, then with it a point will pass about a fixed point in a unit of time through an angle whose measure is ω, and in a time t through an angle whose measure is ωt.

66. Also, as in Art. **16**, if ω be the measure of any particular velocity with certain units of angle and time, then $\dfrac{b\omega}{a}$ is the measure of the same velocity with units of angle and time a and b times the former.

46

As in theoretical mathematics we usually adopt the circular system for measuring angles, and a second for our unit of time, it generally comes to this, that the angular velocity, with which a point will pass in a second through a unit of circular measure about a fixed point, is the unit for angular velocities.

67. PROP. *A point is travelling, in a circle of radius a, with an angular velocity ω about the centre C, and with a linear velocity v along the circumference. To show that v=aω.*

Suppose the point to travel with these velocities for any time t, and let AB be the distance it goes in this time, then

the measure of $AB=vt$, and

the measure of $ACB=\omega t$.

But from Trigonometry

$$AB=a.ACB;$$

$$\therefore v=a\omega;$$

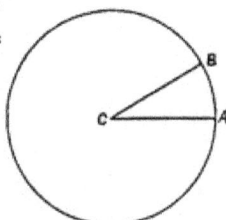

Fig. 13.

or the measure of the angular velocity about $C=\dfrac{v}{a}$.

COR. Generally, let A be any fixed point, P a point moving with a velocity v in a direction making an acute angle ϕ with AP.

Let r be the measure of AP.

Fig. 14.

Then v is equivalent to (1) $v\cos\phi$ along AP, and

(2) $v\sin\phi$ perpendicular to AP.

Now the component (1) has no connection with the angular velocity about A, which is therefore the same as if (1) did not exist and P were moving perpendicularly to AP with the velocity $v\sin\phi$.

Hence the angular velocity of P about $A=\dfrac{v\sin\phi}{r}$.

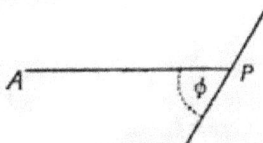

68. *Ex.* A point P is moving in a parabola, with a constant angular velocity about the focus S; show that the linear velocity $\propto SP^{\frac{3}{2}}$.

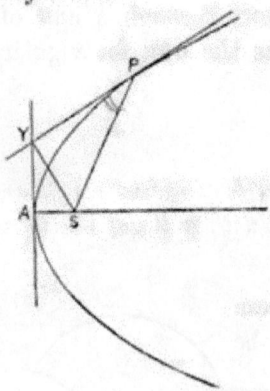

The direction of motion of P is the tangent to the parabola at the position P.

Let A be the vertex of the parabola, SY the perpendicular from S on the tangent at P.

Let ω denote the angular velocity,

v the linear velocity of the point when in any position P.

The component of v perpendicular

to $SP = v.\sin SPY = v.\dfrac{SY}{SP}$;

Fig. 15.

$$\therefore\ \omega = \frac{v.\dfrac{SY}{SP}}{SP} = \frac{v.SY}{SP^2} = \frac{v.\sqrt{AS.SP}}{SP^2} = \frac{v.\sqrt{AS}}{SP^{\frac{3}{2}}};$$

$$\therefore\ v = \frac{\omega}{\sqrt{AS}}\cdot SP^{\frac{3}{2}}.$$

Hence, $\dfrac{\omega}{\sqrt{AS}}$ being a constant, $v \propto SP^{\frac{3}{2}}$.

69. *Normal acceleration.*—Let a point be moving in some plane curve $APQB$, and let P, Q be two adjacent points on AB and PT be the tangent at P.

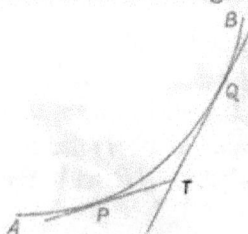

Then PT is the direction of motion at P.

Then, when at P, the point has no velocity perpendicular to PT. Let v be its velocity, and α, β the two components of its acceleration along, and perpendicular to, PT.

Fig. 16.

We call α the tangential, and β the normal, acceleration of the point when at P.

Let t be the time of going from P to Q. Draw QT perpendicular to PT.

Let a', a'' $\begin{cases} \text{be the greatest, and least,} \\ \text{of the components of the} \\ \text{acceleration during } t, \end{cases}$ and β', β'' $\left. \begin{array}{c} \text{parallel,} \\ \text{and perpendicular,} \\ \text{to } PT. \end{array} \right.$

Then a lies between a' and a'',

β „ „ β' „ β''.

Also $PT > vt + \frac{1}{2}a''t^2$ and $< vt + \frac{1}{2}a't^2$,

$QT > \frac{1}{2}\beta''t^2$ and $< \frac{1}{2}\beta't^2$, Art. 62, (2.);

$$\therefore \frac{PT^2}{QT} > \frac{(vt + \frac{1}{2}a''t^2)^2}{\frac{1}{2}\beta't^2} > \frac{2v^2}{\beta'} + \frac{2va''}{\beta'}t + \frac{a''^2}{2\beta'}t^2, \qquad (1),$$

$$\text{and } \frac{PT^2}{QT} < \frac{2v^2}{\beta''} + \frac{2va'}{\beta''}t + \frac{a'^2}{2\beta''}t^2. \qquad (2).$$

Now let t be endlessly decreased, then PT and QT are also endlessly decreased and $\dfrac{PT^2}{QT}$ becomes equal to 2ρ, ρ being the radius of curvature of the curve at P. (Frost's Newton, Art. 78.)

Again, the interval of time being decreased, the values of a', a'' must become closer, and therefore their values must both ultimately become equal to a; similarly β', β'' must both ultimately be equal to β.

Hence the right hand sides of both (1) and (2) must ultimately be equal to $\dfrac{2v^2}{\beta}$; $\therefore 2\rho = \dfrac{2v^2}{\beta}$, or $\beta = \dfrac{v^2}{\rho}$.

Hence the point's motion has at P a normal acceleration, whose measure is $\dfrac{v^2}{\rho}$, wholly independent of any acceleration in the direction of the tangent which the point may happen to have.

70. *Ex.* Let a point be revolving in a circle of radius a with the uniform velocity v. Then at every instant it has an acceleration $\dfrac{v^2}{a}$ towards the centre; and, since its velocity is constant, it has no acceleration in the direction of the tan-

D

gent. Hence this normal acceleration, $\dfrac{v^2}{a}$, is its total, or re-sultant, acceleration. **This is** Newton's Fourth Proposition in Section II.

Further, let ω be the measure of the angular **velocity** about the centre, then $v = a\omega$;

$\therefore$ the measure of the normal acceleration $= \dfrac{a^2\omega^2}{a} = a\omega^2$.

EXAMPLES.—XV.

(1.) If the angular velocity of a wheel 6 feet in diameter, travelling uniformly n miles an hour, be $\dfrac{1}{2n}$, find the unit of time.

(2.) A point is moving along a straight line. Determine the connection between its linear velocity **and its** angular velocity about a fixed point when at a given distance from that point.

(3.) A point is moving in **an** ellipse with uniform angular velocity about the centre. Determine the velocity when in any position, and the normal acceleration.

(4.) A point is moving with uniform linear velocity along the circumference of a circle. Determine its angular velocity about a point in the circumference.

(5.) A point is moving in a parabola with uniform angular velocity about the focus. Determine its angular velocity about the vertex, and the normal acceleration at any point.

(6.) On a straight railway, on the side of which is a line of telegraph posts at intervals equal to their distance from the rail, a train passes a post every second. Show that its angular velocity about any post is $\dfrac{1}{1+t^2}$, where t is the time since passing it.

(7.) A point moves in a circle with uniform velocity. Determine the velocity with which it is approaching a given point in the circle at a given instant, and also the acceleration of its motion towards that point at that instant.

(8.) A is a fixed point in a circle on which the point P moves with uniform velocity. Show that the apparent angular velocity of P about A is constant, and is equal to one-half its angular velocity about the centre of the circle.

V.—RELATIVE MOTION.

71. Def. A point is said to be *in motion relatively* to another when either its distance from the second, or the direction of the line joining it to the second, or both, are being changed.

72. *To judge of the motion of a point (P) relatively to another (Q) at any instant.*

Let A, B be the positions of the two points at any given instant, and suppose that their velocities are then such that, if they remained the same for a time t, the points would move over Aa, Bb.

Produce AB, ab (if necessary) to meet in C. Then the motion of P relatively to Q

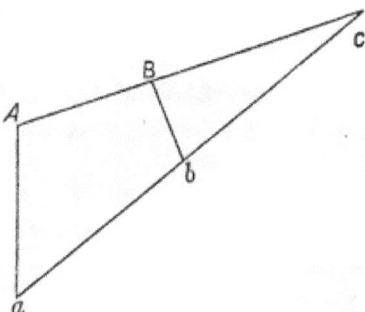

Fig. 17.

at the given instant is such that, if it continued for a time t, the distance of P from Q would be changed by $AB-ab$, and the angle between the line joining them, by ACb.

If ab is parallel to AB, the angle ACb is zero and A is said to have no angular motion, at the instant under consideration, relatively to B.

It will be observed that the motion of B relatively to A is the same in magnitude as that of A relatively to B, only in the opposite direction.

51

73. Prop. *The Relative Motions of a number of Points, at any instant, are not altered by impressing, on each, any the same additional velocity.*

For consider two points, P, Q.

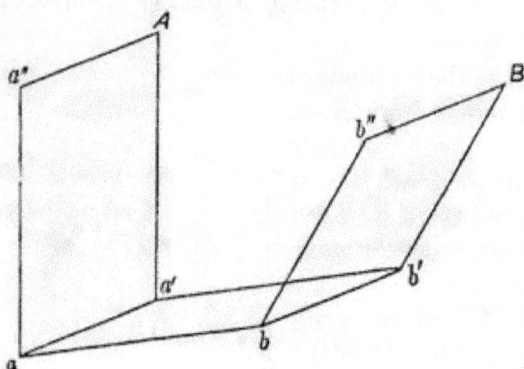

Fig. 18.

At any given instant let the positions of P and Q be A and B, and their velocities such that, if they remained the same for any time t, P and Q would move to a' and b'.

So that $AB - a'b'$ and the angle between the directions of AB and $a'b'$ would represent the change in their relative positions after the time t.

Next, at the given instant let equal velocities be impressed on P and Q in the same direction, and suppose they are such that with them alone P and Q would move over Aa'', Bb'' in time t, so that Aa'', Bb'' are equal and parallel.

Complete the parallelograms $Aa'aa''$, $Bb'bb''$. Then P and Q would now move over Aa, Bb in the time t (Art. 20); and $AB - ab$ and the angle between the directions of AB and ab would now represent the change in their relative positions. But Aa'', Bb'' are equal and parallel; and aa' is equal and parallel to Aa'', and bb' to Bb''; therefore bb', aa', and therefore ab, $a'b'$, are equal and parallel. Hence in both cases the change in the relative positions of P and Q, in the interval t, would be the same. Hence (Art. 72) the relative motions of P and Q at the given instant are the same in both cases, *i.e.* they are not altered by the same velocity being impressed on each.

COR. If the additional velocity be equal and opposite to the velocity of one of them, that point is at rest for the instant.

Hence if we require the motions of any points relatively to any other moving point P during any interval, we suppose that at each instant every point has an additional velocity equal and opposite to the velocity of P at that instant. P may now be considered as fixed, and the motions of all the rest may be calculated by any method which we may have for determining the motions of points relatively to a fixed point.

74. When P is moving during any interval with a varying velocity, then the rate of variation, or in other words the acceleration, of the additional velocity, which each point is to be supposed to have, is equal and opposite to the acceleration of P's velocity. For the additional velocity must vary in the same way as P's, *i.e.* it must have an equal acceleration. Hence, to suppose that at each instant every point possesses an additional velocity equal and opposite to that of P at that instant, is the same thing as to suppose that throughout the interval every point has an additional velocity equal and opposite to that of P at the beginning of the interval and an additional acceleration at every instant equal and opposite to that of P at that instant.

If we suppose that the additional acceleration *only*, and not the additional initial velocity, is possessed by each point, P may be considered as moving in a straight line with a uniform velocity equal to what it has at the beginning of the interval.

75. *Ex. P* is a point moving on a circle, of radius *a*, with uniform angular velocity ω. *Q* is a point moving on a tangent to

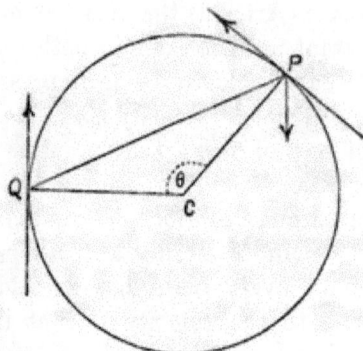

Fig. 19.

the circle with uniform velocity *v*. Find the angular velocity of *P* relatively to *Q* when the latter is at the point of contact.

Let *C* be the centre of the circle, *θ* the angle *QCP*.

Impress on *Q* and on *P* the velocity *v* opposite to *Q*'s given direction, and therefore perpendicular to *QC*; then *Q* is at rest.

The component of *P*'s velocity perpendicular to *QP* is represented by

$$a\omega \cos QPC - v \cos CQP = (a\omega - v) \sin\frac{\theta}{2}.$$

Also *QP* is represented by $2a \sin\frac{\theta}{2}$.

Hence the angular velocity required is represented by $\dfrac{a\omega - v}{2a}$, and is therefore the same wherever *P* may be.

Note.—The angular velocity of *P* about the point of contact is represented by

$$\frac{a\omega \cos QPC}{2a \sin\dfrac{\theta}{2}} = \frac{\omega}{2}.$$

Hence the ratio of the angular velocity about the point of contact to the angular velocity about *Q* when moving through the point of contact

$$= a\omega : a\omega - v.$$

EXAMPLES.—XVI.

(1.) The ends of the hands of a watch are at equal distances from the axis of rotation, and move in the same plane. Determine the relative motion of the two points at any time between 12 and 1 o'clock.

(2.) A point P describes a circle, radius $4a$, with uniform angular velocity ω about a fixed point O, and another point Q describes a circle, radius a, with angular velocity 2ω about P. Show that the acceleration of Q is always proportional to the distance of P from a fixed point.

(3.) Explain why a man walking in a shower of rain in general holds his umbrella a little in front of him.

(4.) Two particles move with constant accelerations in given straight lines. If their velocities when they are in any given positions be given, determine when their relative velocity is least ; and if at any moment their relative velocities in any two directions be in the same ratio as their relative accelerations in the same directions, show that the above relative velocity $=0$.

(5.) Two equal circles touch each other, and from the point of contact two points move on the circles with equal velocities in opposite directions. Prove that one will appear to the other to move on a circle, the radius of which is equal to the diameter of either of the first circles.

(6.) Three equal particles are placed at the corners of an equilateral triangle, and begin to move in their plane with the same angular velocity about their centre of gravity. Compare with this their relative angular velocities.

(7.) A point is moving in an ellipse with a known velocity at a given point, determine its velocity relative to the foot of the normal.

(8.) Prove that, when any points P_1, P_2, ... P_n are moving in any manner, the acceleration of P_n is the resultant of the acceleration of P_n relative to P_{n-1}, of P_{n-1} relative to P_{n-2}, ... and of P_1.

76. Let A, B be simultaneous positions of two points of a figure in motion in one plane, A', B' simultaneous positions of the same two points at some subsequent instant.

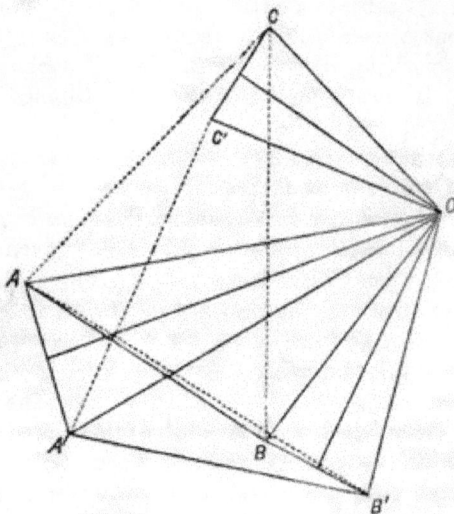

Fig. 20.

Draw lines bisecting AA', BB' at right angles and meeting in O.

Then evidently $AO = A'O$, $BO = B'O$; and since the line, which at one time coincides with AB, afterwards coincides with $A'B'$, we have also $AB = A'B'$; therefore the angles BAO and $B'A'O$ are equal.

Further, consider some third point in the figure and let it be at C when the first two are at A and B, and at C' when they are at A' and B'. Join CO, $C'O$.

Now the triangle, which coincides with ACB at one instant, coincides afterwards with $A'C'B'$; therefore $CA = C'A'$, and the angles BAC, $B'A'C'$ are equal. Then, the angles BAO, $B'A'O$ being equal, the remaining angles OAC, $OA'C'$, are equal. And in the triangles OAC, $OA'C'$, since $OA = OA'$, $AC = A'C'$, and $OAC = OA'C'$; therefore $OC = OC'$.

This result is true wherever C may be.

Thus the second position of each point in the figure is at the same distance from O that the first is. So that we could make the figure move from its first position into its second by turning it round O, keeping this point fixed, each point describing an arc of a circle whose centre is O.

And this is true however near these positions may be to one another.

But, when they become indefinitely near, so that they may be regarded as consecutive positions of the figure, AA', BB', CC', etc. become the directions of motion of the points of the figure in the positions A, B, C, etc. and also coincide with the arcs of circles described round the point which O then occupies. Hence, in any position of the figure, all the points are just then moving in circles having a common centre, called therefore the *instantaneous centre of rotation*, which is for that instant at rest.

We have the following construction for finding the position of this point :—

Find the directions of motion of any two points, A, B, at that instant, and draw AO, BO perpendicular to those directions. The point O, in which these lines meet, is the required point.

Also, if O be joined to any other point, C, of the figure, CO is perpendicular to the direction of motion of C.

77. *Ex.* Let *AB* be a rod sliding with its ends on two straight lines *ZX, ZY*.

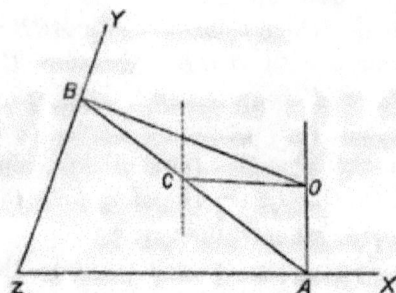

Fig. 21.

Then *ZX, ZY* are the directions of motion of the ends *A* and *B*.

Draw *AO, BO*, perpendicular to *ZX, ZY*, meeting in *O*.

Then *O* is the instantaneous centre of rotation of the rod.

Let *C* be any point in the rod. Join *CO*, then *C* is instantaneously moving perpendicular to *CO*.

78. DEF. A cycloid is the curve traced out by a point in the circumference of a circle, which rolls on a straight line.

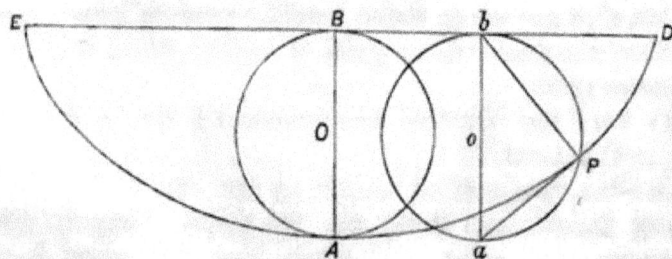

Fig. 22.

Thus let *P* be the point, and *aPb* one of the positions of the circle rolling on the line *DBE*, *ba* being the diameter perpendicular to *DE*.

Suppose that D is the point of contact when P is in DE, and that the circle rolls on, from D, till P comes on to DE again at some point E. Then P traces out the curve DAE, which is said to be an *inverted* cycloid as it is below DBE, which is called the *base*.

A, the position of P when at its greatest distance from DBE, is called the *vertex* of the cycloid and evidently is the middle point of DAE. Also, if AB is drawn perpendicular to DBE, B is the middle point of DBE, and $AB=ab$; also B is the point of contact of the rolling (or *generating*) circle when half way between D and E.

79. The centre of the rolling circle is always at the same distance from DBE, *i.e.* it moves parallel to DBE; $\therefore$ *aob* is perpendicular to the direction of motion of o the centre of aPb; $\therefore$, Art. **76**, the centre of instantaneous rotation is in aob; $\therefore$ b, if moving at all, must be moving perpendicular to aob, *i.e.* along DBE. Now there is supposed to be no sliding of the circle, but only rolling; hence b is not moving along DE; $\therefore$ b is for the instant at rest, *i.e.* it is the instantaneous centre of rotation.

Join bP, aP; then, Art. **76**, P is moving perpendicular to bP, *i.e.* along aP, since bPa, being the angle in a semicircle, is a right angle. Hence Pa is the tangent to the cycloid at P, and Pb the normal.

80. On AB as diameter describe a circle BQA (Fig. 23). It is the position of the generating circle when midway between D and E. It may be called the *auxiliary* circle.

Draw PQN perpendicular to AB, meeting the auxiliary circle in Q and AB in N. P and Q may be called corresponding points on the cycloid and auxiliary.

Then (Figs. 22 and 23) AQ is parallel to Pa, and BQ to bP. Hence AQ and BQ are parallel to the tangent and normal to the cycloid at the point corresponding to Q.

81. As the circle rolls from D to b (Fig. 22), each successive point in the arc bP comes in contact with a point in bD, hence the arc $bP = bD$. Similarly BD is equal to the semicircumference of the circle, and therefore the arc aP is equal to Bb. Also DE is equal to the circumference.

The points D and E are called the cusps.

82. PROP. *The arc AP is double the chord AQ.*

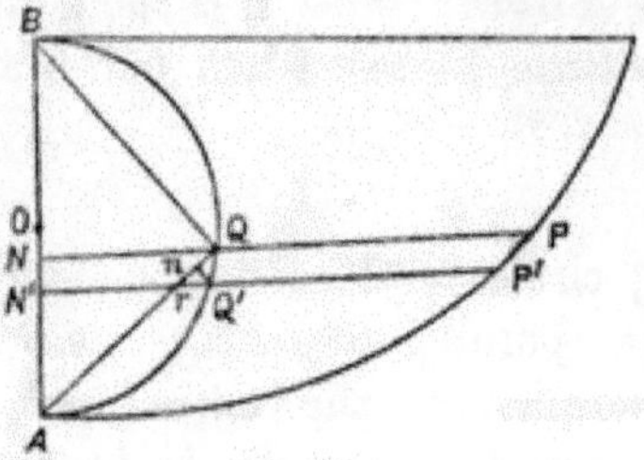

FIG. 23.

Let P' be a point adjacent to P, Q' the corresponding point on the auxiliary circle. Join $P'Q'$ cutting AQ in r. Now ultimately when P' approaches indefinitely near to P, the arc PP' coincides in direction with the tangent at P, *i.e.* is parallel to AQ. Also, QP being parallel to $P'Q'$, $QPP'r$ is ultimately a parallelogram;

$$\therefore rQ = PP'$$

Again, cut off from AQ a distance $An = AQ'$;

$$\therefore nQ = AQ - AQ', \text{ and } \angle AnQ' = \angle AQ'n.$$

Now ultimately the angle QAQ' vanishes,

$$\therefore Q'nA + AQ'n = 2 \text{ right angles};$$

$\therefore AnQ'$ and $AQ'n$ are right angles, and nQ' is perpendicular to rQ.

Ultimately, QQ' coinciding with the tangent at Q to the circle, the angle $Q'QA = \angle QBA$ in alternate segment

$$= \text{complement of } BAQ$$
$$= \angle N'rA$$
$$= \angle QrQ';$$

$\therefore$, ultimately, $rQ' = QQ'$;

$\therefore$ in the isosceles triangle $Q'rQ$ the perpendicular nQ' bisects rQ;

$$\therefore rQ = 2nQ. \text{ Hence } PP' = 2(AQ - AQ').$$

In other words, as P moves from any position to a consecutive one, the increase in the arc AP is double the increase in the chord AQ.

Now let P start from A and move along the cycloid. The arc AP and the chord AQ both start from nothing and continually increase in magnitude. But from above the arc AP increases with double the rate of increase of the chord AQ. Hence when P is in any position the arc AP is double the chord AQ. So in Fig. 22, arc $AP = 2$ chd. aP.

83. Let CD be a semicycloid equal to DA, placed with its vertex at D, touching BD at D, and having its cusp in AB produced. Draw the base Ccd, and the axis Dd.

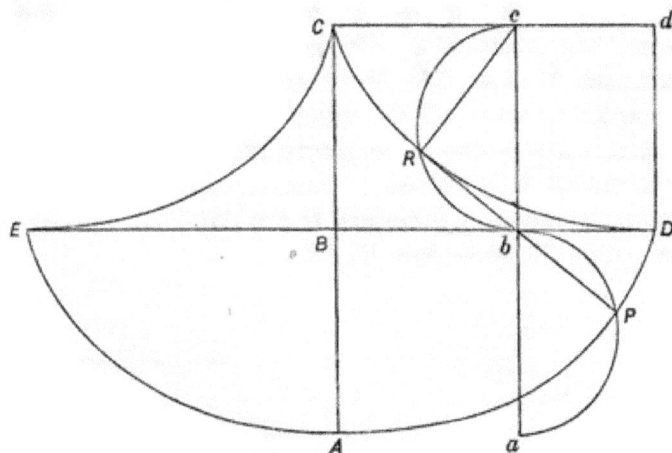

Fig 24.

Let the generating circle of CD start from d, and roll on to C, so as always to touch DB in the same point as the generating circle of DA. Let aPb, cRb be simultaneous positions of these circles; R the point which traces out DC. Join Rb, Pb.

Now R will come at last to C, therefore $Cc = $ arc cR;

$\therefore$ arc $Rb = cd = bD = $ arc bP.

But the two circles cRb, bPa are equal; $\therefore$ chord $Rb = $ chord bP.

Also, $\angle cRb = \angle bPa$, both being right angles; and $bc = ba$;

$\therefore \angle Rbc = \angle Pba$; $\therefore Rb$ and Pb are in one straight line.

Also, since $Rb = bP$, $RP = 2Rb =$ arc RD; and, as a particular case, $CA =$ arc CD.

Also, Rb touches CD at R.

Hence, if CD were a material curve and a string were wrapped round it and gradually unwrapped, the part unwrapped, being kept straight, would lie along the tangent at the point where it left CD, i.e. if R were the point of leaving at any instant, along RP. Also, the part unwrapped $= RD = RP$. Hence the string would reach to P, and the end which was at D would trace out DA as the string unwrapped.

If CE were another material equal semicycloid, having its cusp at C and vertex at E, then as the end of the string was carried past A, along AE, the string would wrap on to CE.

From this property DCE is called the *evolute* of DAE.

Also R is the centre of curvature of the cycloid at P, and RP the radius of curvature.

Hence the radius of curvature at $P = 2bP$.

See Frost's Newton, Art. 75, 76.

VII.—DYNAMICS, MOMENTUM, AND AMOUNT OF FORCE.

84. Up to this point we have treated of the motions of points and figures as matters of pure geometry.

For instance, knowing the velocity of a point we determined how far it would travel in any given time; and so forth.

The science, which we have thus been treating of, namely, the Geometry of a Moving Point's Motion, is called Kinematics. It is a branch of pure Mathematics, since in it we have nothing to do with how motion is produced in physical bodies, but only, taking the fact that a point, or any geometrical figure, has certain rates of motion, etc., we have to determine the relations of the motions to one another and to the spaces traversed in various times.

We shall now pass on to consider the *causes* of motion in bodies, and the effects of such causes. When we have so done, and have found the consequent velocities and accelerations of bodies, we shall be able to apply our kinematical formulæ to determine the spaces traversed in various times. Or we may reverse the process. From knowing the connections between the spaces and times we may determine the motions of the bodies, and then from these ascend and determine something of the laws which regulate the causes of these motions.

85. Dynamics is the science which treats of the application of Force to Matter. It is divided into two parts.

1°. *Statics*, wherein we treat of the application of force to matter so as produce *rest*.

2°. *Kinetics*, wherein we treat of the application of force to matter so as produce *motion*.

86. We must warn the student that the above nomenclature, although now becoming general, is not yet everywhere employed. Thus some works on this subject employ the term Mechanics where we use Dynamics, and Dynamics where we use Kinetics.

87. In this chapter we shall introduce a few physical ideas and terms, which are necessary for the enunciation and comprehension of the Laws of Motion to be given in the next chapter.

88. The student is supposed to have some idea of what is meant by the term Mass, as in the phrase "Mass of a Body." The term Quantity, or Amount, of Matter is equivalent to it. Thus we say that the Quantity of Matter in a body is the Mass of the body.

89. If all the particles of a body, or system, are moving with the same velocity, that velocity is called *the* velocity of the body, or system.

90. *Momentum, or Quantity of Motion.*—We will now offer some explanation of what is meant by the term Momentum.

Consider (1) one ball moving with a certain velocity;

(2) two balls, each of them equal in mass to the first and moving with a velocity equal to that of the first; so that (2) is a system of double the mass of (1).

Then we should say that the quantity of motion of the second system is double that of the first; and we should say the same thing if the two balls in (2) were joined so as to form one body.

And, generally, we say that the quantity of motion of a body varies as the mass when the velocity is constant.

Again, consider two bodies of equal mass, one of which is moving twice as fast as the other in the same direction.

We should say that the quantity of motion of the one was double that of the other.

And, generally, we say that the quantity of motion varies as the velocity when the mass is constant.

The quantity of motion of a system is called its *momentum.*

Thus we see that the momentum is a property of a system the magnitude of which varies,

      (I) as the mass when the velocity is constant,

and (II) as the velocity when the mass is constant.

91. Let m be the measure of the mass of a system;

    v ,,   ,,   velocity of the system.

Then, the momentum $\propto m$ when v is constant, by I,

      and $\propto v$,, m ,,   by II;

        $\therefore \propto mv$ when both m and v vary.

          (Hamblin Smith's *Algebra*, Art. 369.)

[In other words, if m' be the measure of the mass of another system, and v' the measure of its velocity, the momentum of the system we are considering : the momentum of the second system $= m.v : m'.v'$.]

We take as our unit of momentum the momentum of a unit of mass moving with a unit of velocity.

Therefore, the momentum : unit of momentum $= m.v : 1$;

    $\therefore$ the momentum $= m.v$. (unit of momentum) ;

    $\therefore$ the measure of the momentum $= m.v$.

So that denoting the measure of the momentum by M, we have

$$M = m.v.$$

E

92. PROP. *If the various parts of a system are moving in the same direction, the momentum of the system is the sum of the momenta of its parts.*

Let M_1, M_2, etc., be the measures of the momenta of the parts.

Then M_1 represents also the momentum of a body, of mass M_1, moving in the given direction with a velocity 1.

Similarly for M_2, etc.

Hence the momentum of the whole is the same as that of a system, the masses of whose parts are M_1, M_2, etc., and of which every part is moving in the given direction with a velocity 1; *i.e.* it is the same as that of a system, whose mass is $M_1 + M_2 +$ etc., moving with a velocity 1; therefore it is represented by $(M_1 + M_2 +$ etc.$).1$, which $= M_1 + M_2 +$ etc.

93. When different parts of a system are moving in different directions, we **cannot** speak of *the* momentum of the system.

Let m_1, m_2, etc. represent the masses of the parts of the system,

and u_1, u_2, etc. represent the resolved parts of the velocities of m_1, m_2, etc., in *some given direction*.

Then m_1u_1, m_2u_2, etc. represent the momenta of the parts in the given direction; and therefore $m_1u_1 + m_2u_2 +$ etc. represents the momentum of the system *in that direction.*

EXAMPLES.—XVII.

(1.) What momentum is acquired by a mass 3 falling (1) for $3''$, (2) through 100 feet ?

(2.) A body of momentum 10 is moving with velocity 2. What is the measure of its mass ?

(3.) A body of momentum 27 has a mass 9. What is its velocity ?

(4.) A body of mass 5 is travelling with a velocity of 100 yards a minute. What is its momentum, a foot and a second being units of space and time ?

(5.) A body, moving with a momentum 35, traverses 7 feet in 5″. What is its mass?

(6.) Two balls of masses 2 and 3 are travelling eastwards with velocities 5 and 4 respectively, and a ball of mass 11 is travelling westwards with velocity 2. What is the momentum of the whole system of three balls?

(7.) Bodies of masses 2, 3, 4 are moving towards the east, north, and north-west with velocities 9, 12, 16 respectively. Find the momentum of the system (1) in the easterly direction, (2) in the northern direction.

(8.) A ball of mass 10 travels with the acceleration 9 feet per second per second. At one instant its velocity is 3. What is its momentum a minute afterwards?

(9.) A ball of mass 20 has an acceleration 3 in the direction of its motion. By how much does its momentum change in 5 units of time?

(10.) What is the change of momentum of a body in a minute, when its mass is 5 and its acceleration 6, a second being the unit of time?

94. *Change of Momentum.*—Let there be a body, of mass m, in motion. Suppose that its velocity can be represented, at one instant by AB, and at another by CD.

Draw AE parallel, and equal, to CD, and complete the parallelogram $ABEF$.

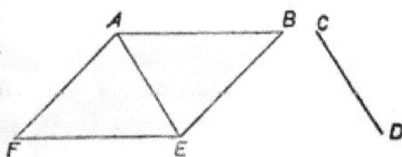
Fig. 25.

Then AF represents the change in the body's velocity in the interval between the two instants.

Let x, y, z be the measures of AB, CD, AF.

Then mx in direction AB, and my in direction CD represent the momenta of the body at the two instants, whilst mz in direction AF represents the *change* of the momentum of the body in the interval; and the momentum of the system is said to be changed by mz in the direction AF in the interval.

95. *Amount of Force.*

(1.) Suppose we had a small body falling, *in vacuo*, from rest for one second. We know, as explained in Art. 59, that it would acquire a velocity 32·2. (A foot and a second being units of space and time.)

(2.) Suppose we had the same body placed on a smooth horizontal plane, and an elastic string attached to it and fixed to some point, so that, when the string is stretched out and let go, the body will move along the plane without rotating. Now, if we attached strings of different elasticities and stretched them to different lengths, we could make the body acquire different velocities within any given interval of time. Suppose then we took a string of sufficient elasticity and stretched it so much that, in the first third of a second after it is let go, the body has acquired a velocity 32·2.

(3.) Suppose we had again the same body at rest, and we struck it a sharp blow. The blow would take up some interval of time so short as to be hardly appreciable. After the blow is over the body would be moving with some considerable velocity, and by properly adjusting the blow we can make it move off with any velocity we choose. Suppose then we make it begin to move with a velocity 32·2.

Here we have instances of motion communicated to matter by different means, namely, the attraction of the earth in (1), the tension of the string in (2), and the pressure of the blow in (3).

In each case we have the same amount of velocity communicated to the same amount of matter, and therefore we say that an equal *amount of force* has been applied in each case.

We took care that this velocity should be communicated in times of different lengths, to bring out the fact that the idea of the amount of force expended is not connected with the length of the time taken to produce the velocity.

96. DEF. By two amounts of force being equal or the same, we mean that they can produce equal amounts of velocity in the *same* body, whatever may be the lengths of the times taken for the purpose.

[Two masses are considered to be equal when the same amount of force will produce in each the same velocity.]

Further, we say that one amount of force is double another when the mass of a body, in which it will produce any particular velocity, is double the mass of a body, in which that other will produce an equal velocity. We justify this statement as follows :—

Suppose the first body to be divided into two parts of equal mass. Each of these halves will require the same amount of force that the second body does, in order to generate in it any the same velocity as in the second, and therefore both these halves together, *i.e.* the whole first body, will require double this amount of force.

Generally, we can say that amounts of force are proportional to the masses in which they produce equal velocities.

97. WE shall give in this chapter the fundamental principles of Dynamics. They are three in number, and are called the Laws of Motion. The first two were recognised by Galileo and his successors; and the third was distinctly enunciated first by Newton.

The forms in which we state them are translations of those given by Newton in his *Principia.*

No rigorous proof of these laws can be given to the student at the outset. All we can do is to give, in some cases, a few simple experiments, which, as far as they go, suggest the truth of the laws. The real proof lies in the fact that the results of calculations with regard to motions of bodies, however complicated, when worked out from these principles, are found to agree with the results of observations made on the bodies themselves.

98. First Law of Motion.—*Every body continues in its own state of rest, or of uniform motion in a straight line, except in so far as it may be compelled by external forces to change that state.*

99. Facts suggesting the Law:—

(1.) "Daily observation makes it appear to us, that any body, which we once see at rest, never puts itself into fresh motion; but continues always in the same place, till removed by some power applied to it."[1]

---

[1] Pemberton's *View of Newton's Philosophy*, London, 1728. Whence also much that is set down in this chapter has been derived.

(2.) If a stone is thrown along ice, the *smoother* the ice is the farther it will go in a straight line. Now the only force acting on it is the roughness of the ice, and we see that the smaller this becomes the less change there is in any given time, leading us to suppose, that, if there were no roughness, there would never be any change in the motion, and that what change there is, is just as much as the roughness is capable of producing.

(3.) If a carriage is suddenly stopped, the passengers appear to be shot forwards. Now, since they are not rigidly attached to the carriage, we suppose that the force, which stopped *it*, has not to a sufficient extent been applied to *them*, and that therefore their motion is in some degree preserved.

(4.) If the carriage be suddenly turned round a corner, the passengers have a tendency to keep their old *directions*. This suggests that, unless the force acts to a sufficient extent on them as well as on the carriage, the direction of their motion will be, more or less, preserved.

100. This law states, that no change in the state of rest or motion of a body will take place, unless there be applied to the body a cause, or force, arising from some source external to the body; and that, where there is a change, whether it be in the direction, or in the magnitude, of the motion, an adequate amount of external force has acted, and the change is just as great as this amount is adequate to produce.

For example, take the following case in addition to those already given :—

When a top is spun it goes on rotating uniformly, except *in so far as* it is retarded by the frictions of the air around it and of the surface supporting it; and by diminishing these we may increase the time it will rotate.

We are thus led to suppose that, if the top were spun *in
vacuo* on a smooth surface, it would always go on rotating
uniformly.

Again, consider the motion of any particle of the top. It
moves in a circle, so that a change in the direction of its
motion is continually taking place. Now for the cause of this
change we look to the cohesion between it and the surrounding
parts of the top. As much of this force of cohesion being
called into play as is sufficient to make each particle move at
every instant in its own circle, instead of continuing to move
in the straight line which is the tangent to the circle at the
then position of the particle.

In Law II. we are told what force is adequate to the pro-
duction of any given change of motion, whether in direction
or in magnitude.

101. Second Law of Motion.—*Change of momentum is proportional to the amount of external force producing it, and is in the
direction of the line of action of the force.*

102. The truth of this law depends on that of the two
following Facts :—

The effect of the same amount of force is always the same,

(I.) Whether the body, to which it is applied, be previously
at rest, or in motion ;

(II.) Whether it be applied alone, or in conjunction with
other forces.

We shall first give some illustrations of these facts suggesting their truth, adding the geometrical statement of each ; and
after that we shall show how to deduce the law from them.

The student must remember the observation in Art. 97, as
to the ultimate proof of the Three Laws.

Also we suppose in our investigations that the bodies acted
on are particles, or, if of finite size, are so acted on that no
rotation is produced. The student must defer for the present
the consideration of rotating bodies.

103. I. That the same amount of force always has the same effect, whether the body be previously at rest or in motion, may be suggested by the following considerations:—

(1.) If a certain amount of force be applied to a body apparently at rest on the earth's surface, the effect will be the same, in whatever direction the force acts, and whatever position on the earth the body has.

Now any body, apparently at rest on the earth, is always being carried forward in the same direction, but with different velocities according to its position on the earth. Yet, as we have said, the effect of the same amount of force is always the same in magnitude, and therefore we cannot doubt but that it would be the same, if the body were absolutely at rest.

(2.) Again, any moveable body on board ship is as easily moved in any direction, whether the ship be at anchor or in steady motion. Thus, if a ball be projected with any force along the deck of the ship, it will go, relatively to the ship, just as far and in the same direction, whether the ship be moving or not. Now the motion relatively to the ship is due to the force of projection alone. Hence we see that the motion which it has in common with the ship does not affect, either in direction or in magnitude, the motion produced by the force.

(3.) A stone, dropped from the top of a mast of a ship in motion, is found to fall at the foot of the mast.

Now, if it had not been dropped, it would have been carried forward through a certain horizontal distance, and, when it is dropped, it is carried forward in the horizontal direction through an equal distance.

This shows that the effect of gravity has not altered the horizontal motion of the stone, and therefore (Art. 27, *ad finem*) whatever change of motion gravity may have produced is in the vertical direction, which is the direction in which gravity produces motion in a body previously at rest.

If, moreover, the time of falling could be noted, we should find that it would always be the same, whatever was the horizontal motion of the ship, showing that the magnitude of the effect of gravity is the same, whatever be the previously existing motion of the stone, and is therefore the same as when gravity acts on the stone initially at rest. This latter point will, however, be more simply exhibited in the following way.

(4.) If a number of balls be projected horizontally at the same instant from a platform, with different velocities, they all reach the ground at one instant—one knock only is heard.

Hence gravity has pulled all the balls through the same distances in the same time.

104. The following is the geometrical statement of this Fact.

Let a body be moving with a velocity represented by AB.

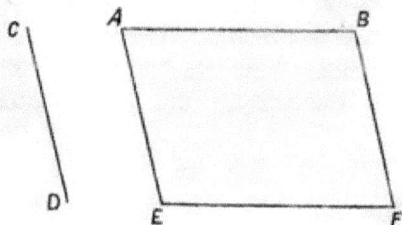

Fig. 26.

Let any single amount of force be now applied to the body.

Let CD represent the velocity which it would produce in the body when initially at rest. Then this fact states that after this application the body's velocity will be changed by the velocity CD; *i.e.* that, as before it was moving with velocity AB alone, it is now moving with velocities AB and CD.

Of course to find the resultant velocity we must compound as in Art. 20, *i.e.* draw *AE* equal, and parallel, to *CD*, and complete the parallelogram *EB*. Thus the diagonal *AF* represents the resultant velocity.

As particular cases, observe that, if *CD* is in the direction of *AB*, the resultant velocity is equal to the sum of the two, and, if *CD* is in the opposite direction, it is equal to their difference.

105. II. That the same amount of force will always produce the same effect, whether it be applied alone or in conjunction with other forces, is suggested by the following considerations :—

(1.) A body is placed on the deck of a ship, which is rough, so that the body cannot slide on account of the ship's motion, and the ship is towed by a tug, which always keeps altering the pace of the ship; then it is always exerting a force on the ship, and therefore also on the body. If now we move the body about on deck in any direction, we find that we require the same amount of force to produce the same motion as when the ship is at anchor or moving steadily.

(2.) If balls be let fall, each down one of a number of smooth and exactly equal pipes held vertically and carried forwards horizontally at different and varying rates of motion, it will be found that the time of falling in each of the pipes is the same, and equal to what it would be if the pipes were at rest or if the balls were falling and the pipes not there.

Now, during their fall the balls were subject to the action of gravity and of the pressures of the sides of the pipes, these last being different in magnitude at different instants and on different balls, but always horizontal in direction.

Hence we see that the vertical effect of gravity is the same, whether the balls be acted on by horizontal forces or not.

Other experiments would be too complicated for explanation in this treatise.

106. The following is the geometrical statement of this Fact. Let a body be moving with a velocity represented by *AB*.

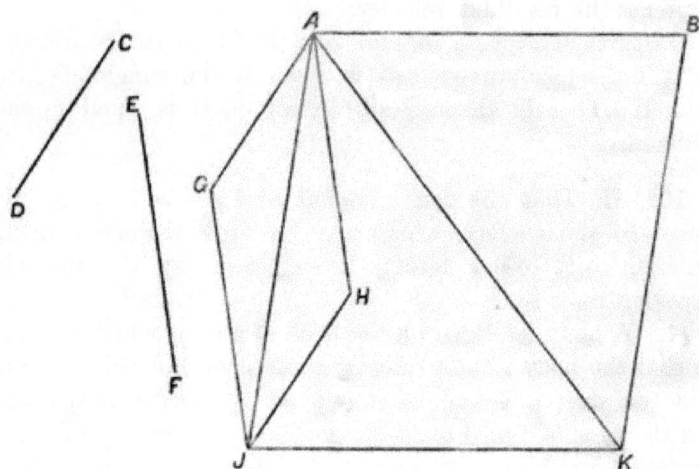

FIG. 27.

Let any two amounts of force be now applied to the body.

Let *CD*, *EF* represent the velocities which they would respectively produce, if each acted alone on the body initially at rest.

Then this fact states that after this application the body's velocity will be changed by the velocities *CD*, *EF*, *i.e.* that, as before it was moving with the velocity *AB* alone, it is now moving with the velocities *AB*, *CD* and *EF*.

To find the resultant velocity, we can compound these together as follows :—

Draw *AG* and *AH* equal, and parallel, to *CD* and *EF* ; complete the parallelogram *GH*. Then the diagonal *AJ* represents the resultant of *CD* and *EF*, and therefore the total change in the body's velocity. Now complete the parallelogram *BJ*. Then the diagonal *AK* represents the resultant velocity of the body.

107. From these two facts we can say that the change of velocity produced by any force on a body, whether it be in motion or not, whether it be acted on by other forces or not, is the same in direction and magnitude as that force would produce if it acted on the body at rest.

Now, first, the direction of the motion produced by a force, acting alone on a body at rest, is called the line of action of the force. Hence one part of Law II. is true, viz., that *the change produced by a force is in the line of action of the force producing it.*

Secondly, let any amount of force be applied to a body at rest, it will produce a certain velocity. After this let an equal amount be applied in the same direction; so that we have now a double amount of force applied. This second amount will produce (by Fact I.) the same change of velocity as if it alone acted on the body at rest, *i.e.* it will produce an amount of velocity equal to what the first did, and in the same direction, which will therefore be added to the velocity produced by the first. Hence we obtain a double velocity with this double amount of force. And, if we again applied an equal amount of force, we should have applied a triple amount of force upon the whole, and should obtain a triple amount of velocity.

Proceeding in this way, we see that the velocity produced by any amount of force is proportional to that amount, when the body is initially at rest, and, therefore also, by Fact I., when the body is initially in motion, and, by Fact II., when other forces are acting. This has been proved by considering the forces always to act on the same body. But the same holds good, if the force acts on different bodies having equal masses; for an amount of force will produce the same velocity in equal masses whether of the same kind of substance or not, Art. **96.**

Now consider bodies of different masses. By the definition of Art. **96** and Facts I. and II. the amount of force varies as the mass, and therefore as the measure of the mass, when the

change in the velocity is constant. And from above, it varies
as the change in the velocity, and therefore as the measure of
the change, when the mass is constant. Hence, when both
vary, the amount of force varies as the product of the measures of the mass and of the change in the velocity, *i.e.* as the
measure of the change of momentum (Art. 94), and therefore
as the change of momentum, or, *the change of momentum is proportional to the amount of force producing it.*

108. *Ex.* Suppose a certain force to act on a body, such that,
after its application, the velocity of the body is changed by a
velocity of 1000 yards an hour. If an equal amount of force
were to act in the same direction on a body of twice the mass
of the former, its change of velocity would be in the same
direction as that of the former, and equal to a velocity of 500
yards an hour.

EXAMLPES.—XVIII.

(1.) A body is travelling with a velocity of 30 feet a second, when a
certain force acts on it, and it subsequently moves with a velocity of
50 feet a second. After that, an amount of force double that of the
former acts on the body. With what velocity will it then be moving?

(2.) Two amounts of force are applied, one to one mass, and the
other to another mass three times as great as the first, and produce
velocities 30 and 20 respectively. What is the ratio between these
two amounts of force?

(3.) A string is stretched twice to the same length, and attached to
two different bodies. At the instant when it becomes slack the
bodies are moving with velocities in the ratio of 3 to 5. Determine
the ratio between their masses.

(4.) If two masses, in the ratio of 3 to 7, are let fall, what is the
ratio between the amounts of force exerted on them by gravity, (1) in
the first second of their fall, (2) in falling through 64 feet?

(5.) A certain force applied to a mass, whose measure is 2, generates
a velocity 5. What velocity will three times this amount of force
generate in a mass 3?

(6.) A body of mass 3 is moving with the velocity 2. What amount of force, acting perpendicularly to the original direction of motion of the body, will be necessary to turn the direction of motion through an angle of 60°?

(7.) Two bodies are moving in parallel directions with the same velocity, and, the same amount of force being applied to each perpendicular to its original line of motion, their directions of motion are changed by 30° and 45° respectively. Compare their masses.

(8.) From a balloon at a given height, and rising vertically with a given velocity, a stone is let fall. Find the velocity with which the stone will strike the earth, neglecting the resistance of the air.

109. Third Law of Motion.—*To an action there is always an equal and opposite reaction ; or, the mutual actions of two bodies are always equal, and directed towards opposite parts.*

110. Illustrations of Law III.

(1.) If anything presses against, pulls, or attracts another, this other presses, pulls, or attracts the first with a force, equal in amount, but in exactly the opposite direction.

(2.) If a load is drawn by a horse, the load reacts on the horse as much as the horse acts upon the load ; for the harness, which is strained between them, presses against the horse as much as against the load, and the momentum of the horse forwards is diminished by as great an amount as that by which the momentum of the load is increased.

Thus, during any interval, let P be the amount of force tending to make the horse move forward, T the amount (exerted by the harness) tending to pull him back. Then this law states that T is the amount exerted by the harness on the load tending to make it move forward. Let F be the amount tending to retard the load. Let m and m' be the masses of the horse and load.

Since, by Law II., any amount of force is proportional to the momentum produced by expending it, we may put the measure of any force $= c$. (the measure of the momentum produced), where c is a constant.

Hence the velocity produced by P in the horse $= \dfrac{P}{cm}$,

,, ,, ,, T ,, $= \dfrac{T}{cm}$,

,, ,, ,, T in the load $= \dfrac{T}{cm'}$,

,, ,, ,, F ,, $= \dfrac{F}{cm'}$;

$\therefore$ the resultant velocity produced during the interval

in the horse $= \dfrac{P}{cm} - \dfrac{T}{cm}$,

and that in the load $= \dfrac{T}{cm'} - \dfrac{F}{cm'}$;

and these are equal ; $\therefore$ $\dfrac{P}{m} - \dfrac{T}{m} = \dfrac{T}{m'} - \dfrac{F}{m'}$;

$$\therefore T(m+m') = P.m' + F.m.$$

If P is just sufficient to keep up the velocity which the horse and load have, the velocity produced during the interval is zero.

Hence, in this case, $\dfrac{P}{m} - \dfrac{T}{m} = 0 = \dfrac{T}{m'} - \dfrac{F}{m'}$; $\therefore P = T = F.$

(3.) Again, when a stone is falling to the ground, the earth attracts the stone with a certain force, and the stone attracts the earth with an equal force.

Thus let M denote the momentum which this force can generate in any given interval of time (t), and m the mass of the stone, m' that of the earth.

Then during any interval, equal to t, of its fall the stone would acquire a velocity $\dfrac{M}{cm}$ towards the earth, and the earth a velocity $\dfrac{M}{cm'}$ towards the stone.

Since the ratio $m' : m$ is so large, the ratio $\dfrac{M}{cm'} : \dfrac{M}{cm}$ is so small, that the velocity acquired by the earth is inappreciable compared with that of the stone.

These instances are sufficient to afford some notion of the meaning of the Law. Its truth cannot be exhibited by any simple experiments. A strong confirmation of it will be found in the agreement of the results of observations on the impacts of bodies with the results of calculations with regard to these impacts, on the assumption that this Law holds, as explained in Chapter XII.

NOTE.—The term *Inertia* is applied to that property of matter, on account of which a body requires, as stated in Law I., some external force to act on it before any motion, or change of motion, can be produced in it.

IX.—MISCELLANEOUS PROPOSITIONS.

111. LET a pair of forces act on a body in directions Ax, Ay, not necessarily for the same time, or for equal times.

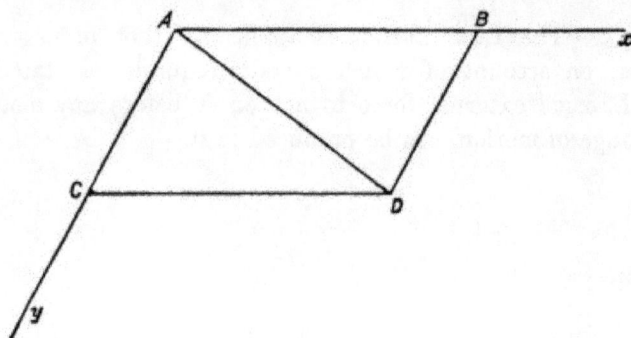

FIG. 23.

Let the velocities which they generate in the body be represented by AB, AC. Complete the parallelogram BC.

Now the diagonal AD represents the resultant velocity, which the body has acquired on account of the action of both forces.

Hence a single force acting in direction AD, and of which the amount expended was such that it would generate the velocity AD, would have the same effect as the two forces of the pair, and is therefore their resultant.

Now since these two forces, and also their resultant, act on the same body, the amounts of force expended by them are proportional to the velocities they generate (Art. **107**).

Hence, on the same scale that the adjacent sides of a parallelogram represent two component amounts of force, the diagonal through their point of intersection represents the resultant amount of force.

Hence amounts of force can be compounded, and resolved, according to the Parallelogram Law, in the same way as forces are in Statics (or, in other words, as the statical effects of forces).

112. We will now give some more extended examples of the application of forces, similar to those in Art. 95 ; especially for the purpose of explaining the meaning of the term *Intensity* of a Force. Take the following cases :—

(1.) A body containing 60 units of mass falling from rest for one second. It acquires a velocity 32·2, and therefore (Art. 91) a momentum 1932.

(2.) A body containing 120 units of mass falling from rest for half a second. It acquires a velocity 16·1 (Art. 59), and therefore a momentum 1932.

(3.) A body containing 40 units of mass pulled by a string, as in case (2.) of Art. 95, of sufficient elasticity, and stretched so much, that in the first third of a second the body acquires a velocity 48·3, and therefore a momentum 1932.

(4.) A body containing 30 units of mass, struck with a sharp blow, such that it acquires a velocity 64·4, and therefore a momentum 1932.

Now, in all these cases, it will be observed that, though different masses acquire different velocities in different times, yet they all acquire the same momentum.

Hence, Art. 95, the same *amount* of force is applied to each.

If the body in (2.) were to go on moving for a whole second from rest, it would acquire a velocity 32·2, and a momentum 3864, and therefore the amount of force applied would be double that in any of the four cases. And if it went on moving for 2″ from rest, it would acquire a velocity 64·4, and a momentum 6928, and now therefore the amount of force would be four times that in any of the four cases.

But now we see that the momenta in (2.) and (3.) have been produced respectively in $\frac{1}{2}$ and $\frac{1}{3}$ of the time, in which an equal momentum was produced in (1.) We say therefore that the *intensities* of the forces acting in (2.) and (3.) are upon the whole 2 and 3 times as great as the intensity of the force acting in (1.)

Also the time taken up by the blow in (4.), though really finite, yet is so inappreciable, and therefore the ratios of its duration to that of the times in (1.), (2.) and (3.) so very small, that we often say that the time of action of such a force is infinitely small, and therefore that the intensity of the force, upon the whole, is infinitely great as compared with the intensities of those in (1.), (2.) and (3.)

113. Again, if we observe the rate at which the velocity in (1.) is generated, we see that equal amounts are generated in equal intervals of time, however small, and in whatever part of the fall they may be taken; and the same is true of (2.) We express this fact by saying that the force acting in (1.) and (2.), namely the force of attraction of the earth, is uniform, or that its *intensity* is always constant. Thus the force in (1.) has one uniform intensity, and the force in (2.) another; and further, we should say that the intensity of the force in (2.) was twice as great as the intensity of that in (1.)

But now, if we observe how the velocity is generated in (3.), we shall see that it is produced much more quickly at the beginning of the motion than at the end. Thus, if we could detect the changes of velocity in two intervals, each $\frac{1}{100}$ of a second long, one near the beginning, and one near the end of the third of a second, we should find that the velocity produced in the first interval is much greater than that produced in the other. Here we should say that the force acting, namely, the tension of the string, is variable, and that its *intensity* varies. Therefore we cannot speak of *the* intensity of the force, as in (1.) and (2.) ; but we must speak of the intensity of the force, at some one instant of the time of motion, or at some one point in the path of the body while the motion is being communicated.

Further, if we think of doing the like with (4.), we find that the whole time of the force's action is so extremely short that we can hardly even conceive in our minds any division of it; so that we cannot distinguish between one instant and another in it; but we must look upon the whole as almost instantaneous.

Also, the body has no time to move with the velocity it acquires at the beginning of the blow before the whole blow is completed, and thus we can hardly conceive of any change of position taking place during the blow, and we cannot speak of the path of the body while the motion is being communicated to it.

So that, although the intensity of the blow probably goes very rapidly through very great variations, and is much less just at the beginning and end than in the middle, yet we cannot speak of the *intensity at any instant;* but must content ourselves with finding the *amount* of force expended during the whole blow. And indeed we do not need to know the intensity at any instant, for there is no question of how far the body is moving during the blow, but only how fast and with what momentum it is moving after the blow is over.

The student will now understand the saying, that, the intensity of a force is the degree of rapidity with which it can generate momentum in bodies.

114. Forces, such as those in (1.), (2.) and (3.), which require a finite time to generate a finite momentum, are called *Finite* Forces.

Forces, such as that in (4.), which require an infinitely

short time to generate a finite momentum, are called *Impulsive* Forces, or *Impulses*.

Probably, as we have indicated, there are no forces in nature which can accurately be called impulses. Yet with a large class of forces, with which we meet, we make no appreciable error by discussing them as if they were impulses, *i.e.* by neglecting the time they take in propagating the motion, as well as the space through which the body acted on has moved during that time, and by looking only to the whole change in momentum produced by them.

When a force is spoken of it is generally understood that a finite force is meant, unless an impulsive force is specially mentioned.

115. By the intensity of a force at any instant being twice as great as that of another, we understand that, if the intensity of each force remains constant for any the same length of time, the momentum generated by the first is twice as great as that generated by the second ; and, generally, we say that, the intensity of a force at any instant is proportional to the momentum which the force can generate in some given interval, supposing it to retain this intensity throughout the interval.

Hence also it is proportional to the *amount* of force then expended during the interval, *i.e.* to the amount of force required to produce the momentum.

Hence, by Art. 111, intensities of forces can be compounded, and resolved, according to the Parallelogram Law.

We shall afterwards show that the intensity of a force is proportional to its statical effect, *i.e.* that a force, which is double some other according to the statical test, will in any the same time produce double the momentum that the other will.

When a force of a given intensity is spoken of, we mean that the force retains the same intensity for the whole time of its action.

116. The following case may tend to explain the fact that there is no line to be drawn between impulsive and finite forces in kind, that the difference is only one of degree; in other words, that an impulse is only a finite force, whose duration of action is finite, though of inappreciable length.

In case (3.) of Art. **112,** if we successively applied to the body strings of greater and greater elasticity, and stretched them more and more, we should find that the required momentum could be produced in shorter and shorter intervals of time, *i.e.* they would become of less and less appreciable lengths. It would seem, therefore, that, by taking strings of sufficient elasticity, there need be no limit to the diminution of the length of time required to produce any given momentum; and, therefore, no limit to the diminution of the errors we should make in regarding the time of action of very strongly elastic strings as practically zero, and the intensities of the tensions while they last as infinitely great.

117. In Art. **113** it was stated that the intensity of a force is uniform when equal velocities are generated in a body in equal times. Hence a force of uniform intensity produces a uniform acceleration in the motion of the body which it acts on.

118. *Ex.* Two forces, *A* and *B*, of uniform intensities, act as follows :—

A acts on a mass 30 for 5 seconds and generates a velocity 9,

B „ 40 „ 2 „ „ 10.

Determine the ratio between the amounts of force expended, and compare the intensities of the two forces.

1°. The momentum generated by *A* is $30 \times 9 = 270$,

„ „ *B* „ $40 \times 10 = 400$;

therefore the ratio between the amounts of force expended

$$= 270 : 400,$$
$$= 27 : 40.$$

2°. A generates a velocity $\frac{9}{5}$ in a second, and therefore a momentum $30 \times \frac{9}{5} (=54)$ in a second,

B generates a velocity $\frac{10}{2}$ in a second, and therefore a momentum $40 \times \frac{10}{2} (=200)$ in a second ;

∴ the intensity of A : the intensity of $B = 54 : 200 = 27 : 100$.

119. By the term Magnitude of a force, we mean, the intensity of the force when we are speaking of Finite Forces, and the amount of force exerted when we are speaking of Impulses.

Instead of the phrase " magnitude of a force," we often, for shortness, use that of " a force." Thus by a certain finite force we mean " a finite force of certain specified intensity ;" and by a certain impulse we mean " an impulse of certain specified amount of force."

EXAMPLES.—XIX.

(1.) If a force act on a body for 3 seconds and generates a velocity 60, determine the acceleration. What would be the acceleration which this force could generate in another body of double the mass?

(2.) Show that the intensity of gravity when acting on a body varies as its mass.

(3.) Two bodies of masses $3m$ and $5m$ are acted on by forces which produce in their motions accelerations 7 and 9 respectively. Compare the intensities of the forces ; and also the amounts of force expended on the two bodies in any the same time.

(4.) Two forces, whose intensities are in the ratio of $3:5$, act on two bodies and communicate velocities 5 and 11 in $3''$. Determine the ratio between the masses of the bodies.

(5.) One force acting on a given mass produces a velocity 100 in $3''$, and a second force acting on another mass produces a velocity 1000 in $3'$. Given that the intensities of the forces are in the ratio of $7 : 11$, determine the ratio between the masses of the bodies, and compare the amounts of force expended on them.

(6.) If two bodies propelled from rest by the same uniform pressure describe the same space, the one in half the time that the other does, compare their final velocities and momenta.

(7.) A force F generates in a body in one minute a velocity of 1300 miles per hour. Which is the greater force, F, or gravity?

120. We can now show that the tests in Statics and in Art. 115 for one force being equal to, or double of, another, agree.

Let there be two forces equal to one another according to the test in Art. 115, so that they generate in equal times equal momenta in bodies initially at rest.

Then, by Art. 91, when applied to the same body, they generate equal velocities in equal times.

Let them be simultaneously applied in opposite directions to a particle. By Art. 105 their joint effect is the difference between the effects they would produce, if applied singly to the particle at rest; but these two effects are equal. Hence their joint effect is *nil, i.e.* they are equal according to the statical test.

Again, let one force be double either of two others according to Art. 115, so that the momentum, which it will generate in any given time in a body initially at rest, is double that generated by either of the two others.

Let the two forces be applied to a particle in one direction, and the one force in the opposite direction. As above, by Art. 105, we can show that their joint effect is *nil*, and therefore the one force is double either of the other two according to the statical test.

Proceeding in this way, we can prove the agreement of the two tests for one force being p times another, *i.e.* the intensity of a force is proportional to its statical effect.

Hence the intensity of one force : the intensity of a second = the statical effect of the first : the statical effect of the second.

121. From the fact that the intensity of a force is proportional to the momentum it can generate *in some given time,* we can obtain an expression for the measure of an intensity.

We may first state the fact as follows :—

Let one force, acting on a mass m for some given time, generate a velocity v; let a second force, acting for the same time on a mass m', generate a velocity v'. Then, the intensities of the forces remaining constant throughout, we have the intensity of the first force : the intensity of the second force

$$=mv : m'v'.$$

Now we take as our unit of intensity that intensity which, if a force retains it for a unit of time, will in that interval generate a unit of velocity in a unit of mass.

Suppose then the above forces to act for a unit of time ; and take the unit of force for the second, and let it act on a unit of mass, then $m'=1$ and $v'=1$. Also take the force, the measure of whose intensity we are going to express, for the first ; then, this being constant, equal velocities are generated in equal times, *i.e.* the acceleration of m is constant. Hence, Art. 39, the measure (α) of this acceleration is equal to the measure (v) of the velocity generated in a unit of time.

Hence the intensity of the force : unit of intensity $=m.\alpha : 1.1$; therefore the intensity of the force $=m\alpha$. (unit of intensity) ; *i.e.* the measure of the intensity $=m\alpha$.

If now we denote this measure by F, we have

$$F=m.\alpha.$$

122. We know that a body falling freely has an acceleration g. Hence, if W is the measure of the weight, *i.e.* the intensity of the earth's attraction on the mass m,

$$W=m.g.$$

123. We obtain an expression for the measure of an impulse from the fact that an amount of force is proportional to the momentum it can generate.

Let one amount of a force be expended on a mass m, and let v be the resulting velocity. Let a second amount be expended on a mass m', and v' be the resulting velocity, then first amount : second amount $= m.v : m'v'$.

We take as our unit of amount that which will generate a unit of velocity in a unit of mass.

Now take this unit for our second amount, and let the mass it acts on be the unit of mass, then $m' = 1$ and $v' = 1$; also take the amount of force, whose measure we wish to express, as the first amount.

Then the amount of force : unit of amount $= m.v : 1.1$;

∴ the amount of force $= m.v$. (unit of amount);

∴ the measure of this amount $= m.v$.

So that, with the above units, the measure of any amount of force is equal to the measure of the momentum it can generate.

Now, by "an impulse" we have said we mean "an amount of force produced by an impulse."

Hence, if an impulse will generate a velocity v in a mass m, we say that the measure (F) of this impulse is $m.v.$, or

$$F = m.v.$$

Also by the unit of impulse we mean such an one as will generate a unit of velocity in a unit of mass.

124. We generally take a foot and a second as our units of space and time, and then our units of velocity and acceleration are determined by Art. 8, 38.

We have not yet stated what force and what mass we intend to take for our units of force and mass; we have only laid down certain conventions as to the relations which shall hold between them, so that when one is settled, the others are determined also.

We generally define our units of mass and force by means of a certain mass defined by Act of Parliament, called the Pound Avoirdupois.

1°. Take as our *unit of finite force* the intensity of the weight of the above mass.

To find the unit of mass. Let *m* be the measure of the above mass, then, the measure of the intensity of its weight being 1,

we have, Art. 122, $1 = m.g.$; $\therefore m = \dfrac{1}{g}$;

i.e. the above mass is $\dfrac{1}{g}$ of the unit of mass.

Hence our *unit of mass* is *g* times the mass of a **pound** avoirdupois.

Then our *unit of impulsive force* is that which can generate a velocity of one foot per second in this unit of mass.

2°. Take as our *unit of mass* the mass of one pound avoirdupois.

To find the unit of force. Let *F* be the measure of the intensity of the weight of this mass, then, the measure of its mass being 1, we have, Art. 122, $F = g$;

i.e. the weight of this mass is *g* times the unit of force.

Hence our *unit of* (finite) *force* is the weight of $\dfrac{1}{g}$ of the mass of a pound avoirdupois.

Of these two systems we generally adopt the first.

125. Further, whichever of the above systems we take, the unit for measuring forces statically is always the force (call it *A*) whose intensity is the unit of intensity.

Now let *f* be the measure of the statical effect of any force (*P*), then statical effect of P : statical effect of $A = f : 1$;

$\therefore$, Art. **120**, intensity of P : intensity of $A = f : 1$;

$\therefore$ intensity of $P = f.$ (intensity of A) ;

$\therefore$ the measure of the intensity of $P = f.$

Hence the statical effect of a force and its intensity are both measured by the same number.

Hence generally when we talk of a force we mean indiscriminately its intensity or its statical effect.

EXAMPLES.—XX.

(1.) What is the mass of a body whose weight is 64 lbs. ?

(2.) What is the weight of a body whose mass is $3\frac{1}{2}$?

(3.) A force 30 acts on a mass 5 for a second ; determine the velocity produced.

(4.) Determine the acceleration produced in the motion of a mass 8 by a force 6.

(5.) What momentum will a force of 6 lbs. produce in $10''$?

(6.) A force of sufficient intensity just to support 5 lbs. acts on a mass 10 ; determine the acceleration produced.

(7.) A force of 12 lbs. acts on a mass of 20 lbs. weight ; determine the acceleration.

(8.) Determine the mass of a body, on which if a force of 5 lbs. act for $12''$, a velocity of 15 feet per second will be produced.

(9.) What force will produce an acceleration 5 in a body whose weight is 5 lbs. ?

(10.) An amount of force 15 is exerted on a mass 3 ; determine the velocity generated.

(11.) What impulse will be required to produce a velocity of 7 in a mass weighing 8 lbs ?

(12.) A weight of 15 lbs. having fallen through 100 feet ; determine the amount of force necessary to stop it.

(13.) When a mass of 20 lbs. weight falls to the ground from a height of 25 feet, determine the shock on the ground.

(14.) In what time will a force, which would support 5 lbs. weight, move a mass of 10 lbs. weight through 50 feet along a smooth horizontal plane, and what will be the velocity acquired ?

(15.) Find the number of inches through which a statical pressure of one oz. constantly exerted will move a mass weighing one pound in half a second.

(16.) In what time will a force, which will just support a weight of 1 lb., move a weight of 6 lbs. through a space of 96 feet on a smooth horizontal table ?

(17.) If a velocity of 100 feet per second is generated by a constant force acting upon a body, weighing 1 lb., in five seconds, what weight would the force statically support ?

(18.) A force, which would support 50 lbs., acts for a minute on a body weighing 300 lbs.; find the velocity it produces.

(19.) A force acting for 10″ produces a velocity of 5 miles an hour in a body weighing 1 cwt. 20 lbs. Find the magnitude of the force in lbs.

(20.) A weight of 20 lbs. lying on a smooth horizontal table is moved by a force equivalent to $2\frac{1}{2}$ lbs.; find the acceleration of its motion.

(21.) What is the magnitude of a blow which will start a weight of 5 lbs. with the velocity 100 feet a second?

(22.) A railway train moving at the rate of 23 miles an hour is brought to rest in 3 minutes, the retarding force of the engine being supposed uniform during the time. Show that this force : weight of the train as 11 : 1890 (assuming g to be 32·2).

(23.) What must be the relation between the magnitudes of the units of time and space, in order that the unit of mass may be the mass of a unit of weight?

(24.) What pressure will be required to produce in 8″ a velocity of 64 feet per second in a weight of 644 lbs.?

(25.) A locomotive 10 tons weight, setting out from rest, acquires a velocity of 20 miles an hour after running through a mile on a horizontal plane, under the action of a constant pressure. Calculate in pounds approximately the difference between the moving and resisting forces.

(26.) A foot and a second being units of space and time, and the unit of acceleration that which generates in a second a velocity of one foot a second, and the mass of a cubic foot of water, which weighs 1000 oz., being unit of mass; find the unit of weight.

(27.) A ball of 10 lbs. weight is held in the hand and lowered with an acceleration (1) of 10 feet per second per second, and (2) of 40 feet per second per second. What are the pressures on the hand in the two cases?

(28.) What is the unit of weight when the unit of mass weighs 5 lbs., and a foot and second are the units of space and time?

(29.) The scale-pans of a pair of scales being first allowed just to rest on a horizontal table, the fulcrum is raised (1) gradually, or (2) with a jerk; of what is it that the relative amount is *immediately* measured in the respective cases?

126. We shall often have to consider a problem similar to, or a particular case of, the following :—

Let a number of finite forces act, in one plane, on a body of mass m; and let their action be such that it moves without rotation.

Draw in the plane two lines, Ox and Oy.

Each force is equivalent to two components, one parallel to Ox and the other to Oy.

Let X and Y denote the algebraic sums of these two sets of components respectively.

Then the system is equivalent to the two forces X and Y.

Let the acceleration of the body's motion, produced by the given system, be equivalent to the two components α and β, parallel to Ox and Oy.

Then X and Y between them must be capable of producing these accelerations. Also, each force produces the change in its own direction, by Law II.; $\therefore X$ produces α, and Y produces β;

$$\therefore, \text{Art. 121, } X = m\alpha, \ Y = m\beta.$$

Generally Ox and Oy are at right angles to one another, and then X and Y are the sums of the *resolved parts* of the forces in the directions Ox and Oy.

127. Similarly, if the forces be impulsive, let X and Y denote the algebraic sums of the resolved parts, in two perpendicular directions, of the amounts exerted by the forces; u and v the *changes* in velocity, in the directions of X and Y, produced by the forces.

Then, as in Art. 126, u is due to X, and v to Y;

$$\therefore, \text{Art. 123, } X = mu, \ Y = mv.$$

Also the above is true, if X and Y are the amounts expended by a system of finite forces, and u and v the consequent changes in the velocity in the two directions.

128. PROP. *The momentum of a* **system** *in any direction is not altered by* **any action** *between the parts of the system.*

On account of this action some amount of force will be exerted by one part on the other. Therefore, by Law III., an exactly equal and opposite amount will be exerted by the second on the first part. Hence the resolved parts of these forces in any given direction will be exactly equal and opposite. Also, by **Law II.**, the change of momentum of each body in this direction will be proportional to the resolved part, in this direction, of the force producing it.

Hence the changes in the momenta of the two parts in this direction will be exactly equal and opposite.

Hence there will be no change in the momentum of the whole system in this direction.   Q.E.D.

129. Suppose a fine inelastic string is moving along a straight line or a curve, into which it is bent by being stretched over a surface.

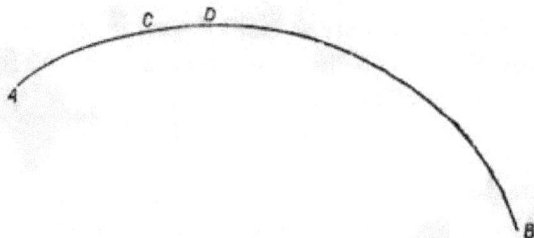

FIG. 29.

Let *ACDB* be the string.

At any point *C* the portion *AC* pulls the portion *CB* with some force along the tangent at *C* from *C* towards *A*.

Hence, by **Law III.**, *CB* pulls *CA* with an *equal* force along the tangent at *C* from *C* towards *B*.

We talk of either of these forces as *the* tension of the string at *C*.

The following is the case which we generally consider:—The surface (if there is any) is smooth, and the string is so fine that the mass of any portion may be neglected, and no other external forces act on any portion but the tensions at the ends, the pressures of the surface at the various points in contact and the weight. We can, in this case, show that the tensions at all points are the same.

Let D be an adjacent point, which we will ultimately make to move up to C. Then CD may be considered as a straight line. The pressure of the surface at every point of CD is perpendicular to its length, and therefore to its direction of motion, and therefore cannot affect its motion in that direction.

Also, since the mass of CD may be neglected, its momentum and weight, which are proportional to its mass, may be neglected.

Hence the only forces which can affect the motion in the direction of its length are the tensions at C and D.

Denote them by T and T', and the acute angles they make with CD by θ and θ'.

Hence, resolving along CD, $T\cos\theta - T'\cos\theta' = 0$.

But ultimately, when D moves up to C, θ and θ' both tend to 0, and therefore $\cos\theta$ and $\cos\theta'$ to 1; $\therefore T = T'$.

Hence the tension at any point is equal to the tension at the succeeding point, and so going on we can show that the tensions at all the points are the same.

Hence we can talk of *the* tension of the string, without specifying the point.

Thus the tension of a tight inelastic string, being the same at all points of it *at the same instant*, will exert on two particles, attached to its ends, *equal* amounts of force in the *same* interval of time.

130. In all instances of motion discussed in this treatise we neglect the effect of the resistance of the air, or, what comes to the same thing, we suppose the motions executed *in vacuo.*

G

131. *Ex.* P and Q are two particles connected by a string, which is passed over a smooth peg; initially, Q rests on a table, on which some of the string is coiled, and P descends freely.

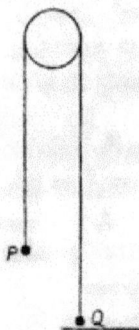

After a time, P has descended so far that the whole of the string becomes stretched.

Then, at one instant Q is at rest on the table, and at the next it is moving upwards with the same velocity that P then has.

Hence Q has acquired a velocity instantaneously, and therefore an impulse must have acted on it by means of the tension of the string.

Let T denote the amount of force exerted by this tension to move Q off.

Fig. 30.

Let V be the velocity of P just before Q has begun to move, V' that of P and Q just after   „   ;
and let m and m' be the masses of P and Q.

Then the momentum $m'V'$ has been produced in Q by T;

$$\therefore T = m'V'.$$

Also, an equal amount of force has been exerted on P (Art. 129), and has produced the change, $mV - mV'$, which has taken place in the momentum of P;

$$\therefore T = mV - mV'.$$

Hence $m'V' = mV - mV'$;

$$\therefore V' = \frac{mV}{m+m'}; \quad \text{and} \quad T = \frac{mm'V}{m+m'}.$$

Also the amount of force expended on the peg is $2T$

$$= \frac{2mm'V}{m+m'}.$$

During the very short time, in which the change of velocity was taking place, the intensity of the force of the impulse went very rapidly through very considerable changes, and at one time was very great indeed. Sometimes the string is not able to bear this great intensity and snaps.

Suppose the string was just able to support a weight of 50 lbs., then the string will break, if during the impulsive action of the string the intensity of the tension exceeds 50 lbs.

132. *Ex.* P, Q, R are three particles. P and Q are connected by a string stretched over a smooth peg; Q and R are connected by a string, part of which, with R, lies on a table vertically below Q. And suppose that P is descending, and Q ascending, when the second string becomes stretched.

Then, at one instant R is at rest, and at the next it is moving upwards with the same velocity that P and Q then have, which is different to what P and Q had at the previous instant.

Hence there has been an instantaneous change in the velocity of each of the particles, and therefore an impulse must have been exerted by each of the strings.

Let T and T' be the amounts of force exerted by the strings PQ and QR, respectively.

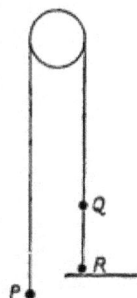

Fig. 31.

Let V be the velocity of P and Q just before R begins to move,

and V' the velocity of P, Q and R just after R begins to move.

Let m, m' and m'' be the masses of P, Q and R.

Then the momentum $m''V'$ has been produced in R by T';

$$\therefore T' = m''V'. \qquad \qquad (1)$$

Also T has produced the change, $mV - mV'$, which has taken place in the momentum of P;

$$\therefore T = mV - mV'. \qquad \qquad (2)$$

And the change, $m'V - m'V'$, in the momentum of Q has been produced by the joint effect of T and T';

$$\therefore T' - T = m'V - m'V'. \qquad \qquad (3)$$

From these equations V', T and T' may be found in terms of V, m, m' and m''.

The amount of force expended on the peg is $2T$.

133. *Ex.* On a smooth horizontal table lies an inelastic string, and to each end is attached a particle. One particle is projected along the table. Having given the velocity of this particle and the inclination of its direction of motion to the string when the latter becomes stretched, determine the impulse along the string and the initial motion of the other particle.

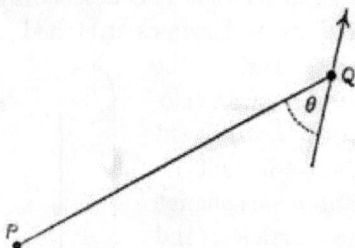

Let P, Q be the particles, m, m' their masses, Q the one projected, V its velocity, and θ the angle its line of motion makes with the string when it becomes stretched.

Let T be the amount of force exerted by the impulse along the string on Q and P; V' the velocity with which P begins to move.

Fig. 32.

The motion of Q just before the impulse is equivalent to $V\cos\theta$ along PQ, and $V\sin\theta$ perpendicular to PQ.

The only force acting is along QP. Hence there is no change in the motion of Q perpendicular to QP.

Also, after the impulse the velocity of Q along PQ is the same as that of P, since the string is inextensible; and the change of Q's momentum, $m'(V\cos\theta - V')$, in the direction of PQ is wholly due to T;

$$\therefore T = m'(V\cos\theta - V').$$

Also the impulse T acting on P produces the momentum mV';

$$\therefore T = mV'.$$

Hence $mV' = m'(V\cos\theta - V')$;

$$\therefore V' = \frac{m'V\cos\theta}{m+m'}; \quad \text{and } T = \frac{mm'V\cos\theta}{m+m'}.$$

Hence Q is moving at the end of the impulse with a velocity, whose components are $\dfrac{m'V\cos\theta}{m+m'}$, along PQ,

and $V\sin\theta$, perpendicular to PQ.

NOTE 1. We have explained in the present Chapter, and in Chapters I. and II., the standards usually adopted. Other systems have also been proposed; notably one in the Report of the British Association for 1873, to which we refer the student. It is proposed in this Report to use the Centimetre where we generally use the foot, and the Gramme instead of the Pound Avoirdupois. [One foot=30.48 centimetres, and 1 lb.=453.6 grammes.] So that the unit of velocity would be that with which a centimetre is traversed in a second. Also the unit of force would be that which could, in one second, generate in a gramme a velocity of one centimetre a second. The other units would be altered in a corresponding manner.

NOTE 2. By Art. 121, the measure of the intensity and the measure of the amount of force expended in a unit of time are the same.

If A is the amount expended by a constant finite force in an interval t, then the amount expended during the unit of time is $\dfrac{A}{t}$, and therefore $\dfrac{A}{t}$ is the intensity.

Also, if A is the amount expended by a blow or other varying force which takes up a time t, then $\dfrac{A}{t}$ is the *average* intensity of the blow and is sometimes *assumed* to be the intensity of the blow throughout.

EXAMPLES.—XXI.

(1.) An engine, whose power is sufficient to generate in $1''$ a velocity of 150 feet a second in a mass M (which is its own mass), is attached to a carriage, mass $= \dfrac{M}{2}$, by means of an inelastic weightless chain 3 feet long; this carriage again is attached in exactly the same way to another, mass $= \dfrac{M}{2^2}$; this to a third, mass $= \dfrac{M}{2^3}$. The engine and carriages are successively in contact when the train starts. Show that the last carriage will begin to move with a velocity of 33 feet per second nearly.

(2.) A, B, C are three equal balls situated at the angular points A, B, C of an equilateral triangle, and connected by two fine strings AB, BC. The ball B receives an impulse in a direction at right angles to AC, and in the plane ABC. Prove that the velocity produced thereby in B is $\frac{2}{3}$ of what it would have been if B had been free.

(3.) A number of equal heavy particles are fastened at equal distances a, on an inelastic string, and placed in contact in a vertical line; show that, if the lowest be then allowed to fall freely, the velocity, with which the nth begins to move, is equal to

$$\sqrt{ag\frac{(n-1)(2n-1)}{3n}}.$$

(4.) A shot of mass m is fired from a gun of mass M with a velocity u relative to the gun; show that the actual velocity of the shot is $\dfrac{Mu}{m+M}$, and that of the gun $\dfrac{mu}{m+M}$.

(5.) Two particles are tied as in Art. **133**, and are moving on a smooth table in a given manner when the string becomes stretched. Determine the impulsive tension and the motion of each particle immediately after the impulse.

(6.) Three particles are tied at different points of a string and placed on a smooth table, so that the string between two of them is stretched and the two particles at rest. The other particle is now projected along the table, and is moving in a given manner when the string becomes tight. Determine the motions of the particles immediately afterwards.

X.—UNIFORMLY ACCELERATING FORCES IN THE LINE OF MOTION.

134. WE shall in this chapter discuss several problems, in which particles are acted on by forces which cause them to move with rectilinear uniformly accelerated motions.

135. Two particles, P and Q, are connected by a string stretched over a smooth fixed peg, each particle hanging freely.

The string always being stretched, P and Q, at every instant, must be moving with equal velocities, one up, and the other down. Also their accelerations must be equal and opposite.

Let m denote the mass of P, m' of Q, and suppose $m > m'$.

FIG. 33.

Let T denote the tension of the string.

Then the forces acting on P are, the weight, mg, downwards, and T upwards,

and the forces acting on Q are, the weight, $m'g$, downwards, and T upwards.

Now T must lie between mg and $m'g$; for if it were greater than each of them, *both* P and Q would have an acceleration upwards, and if less, downwards.

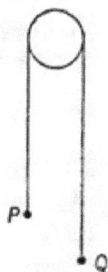

Hence the forces acting on P are equivalent to $mg-T$ downwards,

and the forces acting on Q are equivalent to $T-m'g$ upwards;

$\therefore$ the acceleration of P's motion $=\dfrac{mg-T}{m}$ downwards,

and   ,, Q's   ,, $=\dfrac{T-m'g}{m'}$ upwards.

And these, as we have said, are equal in magnitude;

$$\therefore \frac{mg-T}{m}=\frac{T-m'g}{m'}; \therefore T(m+m')=2mm'g;$$

$$\therefore T=\frac{2mm'g}{m+m'}; \quad \cdot \quad \cdot \quad \cdot \quad (1)$$

$\therefore$ the acceleration (a) of either particle

$$=g-\frac{T}{m}=g-\frac{2m'g}{m+m'}=\frac{m-m'}{m+m'}g, \quad \cdot \quad (2)$$

which is constant.   Hence we can apply the formulæ of Art. **58.**

The following are examples of their application :—

1°. Suppose P and Q to be initially at rest.   To find their velocity v after t seconds.

We have $v=at=\dfrac{m-m'}{m+m'}gt.$

2°. Suppose Q initially projected downwards with a velocity u. To find how soon the system will come to rest.

Let t be the time which elapses before it comes to rest.

Then, at the beginning of the time Q has a velocity u downwards, throughout the time it has an acceleration a upwards, and at the end of the time its velocity is 0 ;

$$\therefore 0=u-at; \therefore t=\frac{u}{a}=\frac{u(m+m')}{(m-m')g}.$$

To find also how far (s) Q has then travelled, we have

$$0=u^2-2as; \therefore s=\frac{u^2}{2g}\left(\frac{m+m'}{m-m'}\right).$$

After this the system begins to move the opposite way.

From (1) we see that T is constant throughout the motion, and is as much as would be required to support a weight $\dfrac{2mm'}{m+m'}g$. Hence the strain on the peg $= 2T = \dfrac{4mm'g}{m+m'}$.

If W, W' are the weights of P and Q, $m = \dfrac{W}{g}$, $m' = \dfrac{W'}{g}$;

$$\therefore \; T = \frac{2 \cdot WW'}{W+W'}, \text{ and } a = \frac{W-W'}{W+W'}g.$$

136. A particle P is placed on a smooth inclined plane.

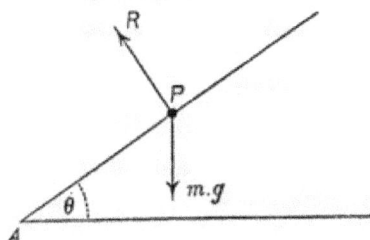

FIG. 34.

[Any line in an inclined plane, perpendicular to the intersection of the plane with a horizontal plane is, of all lines that can be drawn in the former, the one that makes the greatest angle of inclination to the latter, and is therefore called a line of greatest slope. Its inclination to the horizon is the inclination of the plane to the horizon, called the elevation of the plane.

Any vertical plane perpendicular to the intersection of a horizontal plane with the inclined plane cuts the latter in one of its lines of greatest slope.]

Let m denote the mass of P, R the pressure of the plane.

Then the forces acting on P are, the weight, mg, vertically downwards, and R perpendicular to the plane upwards.

Let θ be the elevation of the plane.

Draw PA the line of greatest slope through P; then PA makes an angle $\frac{\pi}{2} - \theta$ with the direction of mg.

Also R and mg both lie in the vertical plane through PA. Hence the forces are equivalent to two,

      (1) $R - mg\cos\theta$, perpendicular to plane,

      (2) $mg\sin\theta$, down PA.

Now, since P remains on the plane, there is never any motion perpendicular to the plane;

      $\therefore$, from (1), $R - mg\cos\theta = 0$;

      $\therefore R = mg\cos\theta$.

Hence the pressure on the plane is always the same.

Again, from (2) we see that P always has an acceleration

down $PA = \dfrac{mg\sin\theta}{m} = g\sin\theta.$.   .   .   .   .   (3)

Hence, if P is initially at rest at some point in PA, or if projected up, or down, PA, it always has a constant acceleration in its line of motion, and we may apply the formulæ of Art. 58.

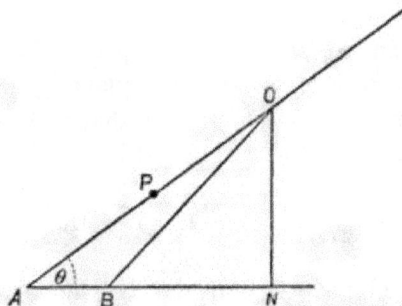

FIG. 35.

1° Suppose P initially at rest at O. To find its velocity, v, at a point A, at a distance s from O.

We have $v^2 = 2.g\sin\theta.s$;

      $\therefore v = \sqrt{2.g\sin\theta.s}.$

Now draw ON perpendicular to the horizontal plane through A, then $ON=s.\sin\theta$; $\therefore v=\sqrt{2g.ON}$, and this is independent of the inclination θ, and depends only on the vertical distance of A below O.   Hence, if OB represented any other inclined plane, down which P were allowed to descend, then the velocity when on a level with A would be the same as at A, and would be the same as if P had fallen freely from O to N.

2° If u is the velocity at O, then $v^2=u^2+2.g\sin\theta.s$
$$=u^2+2g.ON.$$

Hence again the difference between the squares of the velocities at two points depends only on the *vertical* height between them.

3° If P is initially at rest at O, to find the time (t) of descent to A.

We have $s=\frac{1}{2}.g\sin\theta.t^2$; $\therefore t=\sqrt{\dfrac{2s}{g\sin\theta}}$.

137. The same investigations hold good, if, instead of moving on a line of greatest slope of a plane whose elevation is θ, P were a bead moving on a smooth wire, or in a smooth groove, which makes an angle θ with the horizon. In any one of these three cases P would be said to be moving on a *smooth straight line* inclined to the horizon.

138. If the plane is rough, let μ' denote the constant ratio between the Friction and the Pressure during the motion, called the coefficient of Kinetic Friction.

Hence the friction $=\mu'R=\mu'mg\cos\theta$.

Then, when the particle is moving upwards, the friction acts down the plane; therefore the force acting on the particle along PA is $mg\sin\theta+\mu'mg\cos\theta$;

$\therefore$ its acceleration is $g\sin\theta+\mu'g\cos\theta$, downwards.

Similarly, when it is moving downwards, we can show that its acceleration is $g\sin\theta-\mu'g\cos\theta$, downwards.

When the particle starts by moving up the plane, it will eventually come to rest; and if θ is not greater than $\tan^{-1}\mu$, where μ is the coefficient of Statical Friction, it will remain where it comes to rest.

139. Let ABC be a vertical circle, of which A and B are the highest and lowest points.

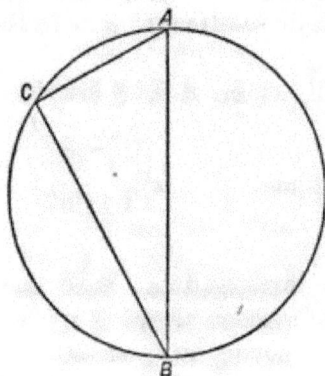

FIG. 36.

Let AC be any chord through A. Then the acceleration of a particle sliding down AC, supposed smooth, is $g\cos CAB$.

Hence the time of descent down AC, from rest, is $\sqrt{\dfrac{2.AC}{g\cos CAB}}$; but $AC = AB\cos CAB$; therefore the time is $\sqrt{\dfrac{2AB}{g}}$, which is the time a particle would take to fall freely from A to B; therefore the times of descent down all chords through the highest point are the same.

Similarly, we can show that the times of descent down all chords, such as CB, through the lowest point are the same.

140. To find the straight line of quickest descent to a vertical circle from a point in its plane.

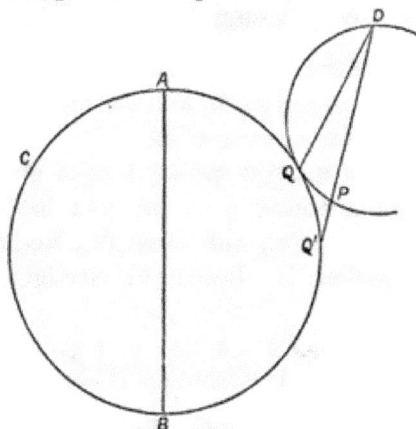

Fig. 37.

Let *ABC* be the circle, *D* the point.

I. Draw the circle which has *D* for its highest point and
touches *ACB* externally. Let *Q* be the point of contact.
Then a particle, starting from rest at *D*, will reach *ACB* in
less time by travelling along *DQ* than by going along any
other straight line.

For let *DPQ'* be any other line through *D*, meeting *ABC* in
Q' and the second circle in *P*.

Now the time of descent down *DP* is equal to the time down
DQ; but the time down *DQ'* is obviously greater than that
down *DP*, and therefore greater than the time down *DQ*.

Hence *DQ* is the line of quickest descent from *D* to the
circle.

II. It can be shown by Geometry that *DQ* produced will
pass through the lowest point of *ABC*. Hence we have the
following simple geometrical construction.

Join the given point to the lowest point of the given circle;
the part outside the circle is the line of quickest descent
required.

141. If we draw a circle, having *D* for its highest point, and touched internally by *ACB*, the line joining *D* to the point of contact is the line of slowest descent from the point to the circle.

142. There have been devised a large number of problems, similar to the above, requiring to find the line of quickest, or slowest, descent from a point to a line.

They all involve a method similar to the above; namely, I., describing a circle touching the line and having the given point for its highest point, and then, II., finding a simple geometrical construction for drawing the straight line without describing the circle.

If the descent is to be from a line to a point, we begin by describing a circle having the point for its lowest point.

143. The results of this chapter may be used to determine the numerical value of *g*.

For instance, in Art. 136, the acceleration $= g \sin\theta$. If then we observe the time (t) of passing over any distance (s) from rest, and the angle (θ) of the inclination of the plane, we can determine g by the formula $s = \frac{1}{2} g \sin\theta . t^2$. This method was proposed by Galileo; but the impossibility of finding a plane sufficiently smooth prevents us from using it.

Also we cannot observe the direct vertical descent of a falling body, since the rapidity of its motion prevents the time of descent through any distance being accurately determined, and also produces too great resistance from the air for the acceleration to be considered uniform. It was for this reason that Galileo proposed the above plan, as with it, by making θ sufficiently small, we can make the acceleration as small as we please.

About the year 1780 Mr. Atwood invented a machine for producing a slow uniformly accelerated descent of a body, the expression for the acceleration involving *g*. (See *A Treatise on the Rectilinear Motion and Rotation of Bodies*; with a description of original experiments relative to the subject. By G. Atwood, M.A., F.R.S., late Fellow of Trin. Coll., Cam., 1784.)

144. A description of this machine is given in Art. 145, which will be the better understood for the following remarks.

In Art. 135, if m and m' are increased by the *same* amount x, then the acceleration is decreased from $\frac{m-m'}{m+m'}g$ to $\frac{m-m'}{m+m'+2x}g$; no change being made in the numerator of the expression.

Again, the peg was taken to be *smooth*.

This would be impossible in practice, and therefore the motion of the particles would be checked by the friction. To obviate this, we may pass the string over a small pulley turning on an axis, whose ends are pivots supported on sockets, or better still, on two pairs of small wheels, called friction wheels.

For then the friction is produced, not by the sliding of the string over the *surface* of the peg, but by the axes of the pulleys, as they rotate, rubbing against their supports across the *lines* of contact, and therefore is so slight that the diminution of the acceleration thus caused may be neglected.

By this means, however, we have introduced a new check to the motion. For the forces acting on the system have to change the motions of the pulleys as well as those of the weights. Hence the rate of change of the motion of these latter is diminished. Its actual amount can be easily calculated by Rigid Dynamics, when the make of the pulleys is known. It is there shown that the acceleration of the weights is the same as if the pulleys were smooth and fixed, and a certain amount $\left(\text{say } \frac{n}{2}\right)$ added to each weight, *i.e.* the acceleration can be represented by $\frac{m-m'}{m+m'+n}g$. We may also determine n by experiment.

145. *Atwood's Machine.*—Two exactly similar and equal bodies P and Q, each of mass m, are connected by a fine thread passing over a pulley, whose pivots are supported by four

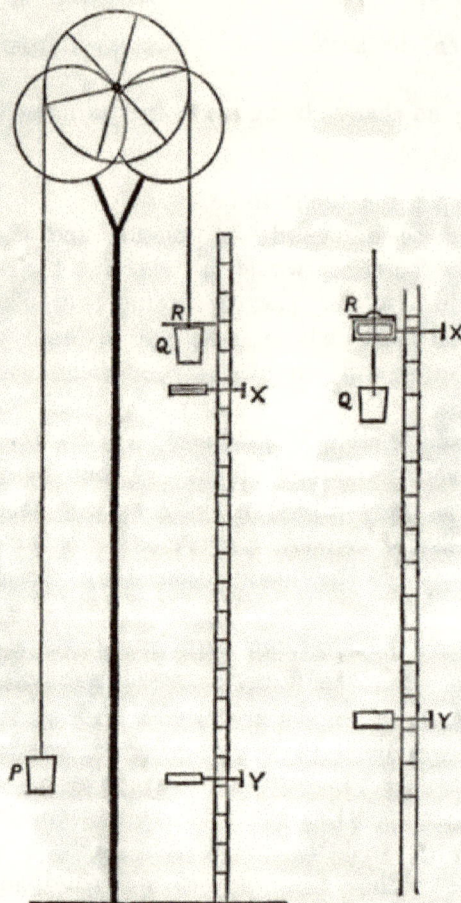

FIG. 38.

friction wheels (two only being shown in the figure) on the top of a pillar, by the side of which is placed a graduated scale so that Q descends close to it. To this scale two moveable frames, or platforms, X and Y, are attached by screws so as to be capable of being screwed to any points of the scale.

The upper frame X is pierced so as to allow Q to pass through it.

Place Q above X. Then P and Q, being equal, will remain at rest. Now let a small rod R, of mass z, be placed on Q, projecting from each side of it, so as not to be able to pass through X. P ascends and Q, with R, descends with uniform acceleration, which would be $\dfrac{z}{2m+z}g$, if the pulleys were fixed and smooth, but which is $\dfrac{z}{2m+z+n}g\ (\equiv f)$, since the pulleys, as well as the masses P, Q and R must have the motion communicated to them.

Atwood shows how n may be determined by experiment.

When Q passes X, R is caught off. P and Q have now no acceleration, and therefore move on, with the velocity they have then acquired, till Q reaches Y.

There is also clock-work attached, by which the times of descent of Q through any spaces may be accurately observed.

Let the system begin to move, and observe the time t which elapses till Q passes X, and the time t' from that moment till it reaches Y, and measure XY (s).

Let v denote the velocity acquired when Q passes X.

Then Q moves over the space s in the time t' with the uniform velocity v;

$$\therefore s=vt'\ ;\ \therefore v \text{ is found.}$$

But, in the time t, Q has moved with the uniform acceleration f, and has acquired the velocity v;

$$\therefore v=ft\ ;\ \therefore f \text{ is found.}$$

But m, m', z, n are known; $\therefore g$ is found.

The value of g is found to be slightly different at different points of the Earth.

EXAMPLES.—XXII.

(1.) Two masses, m_1, m_2, are connected by a string, which passes over a smooth peg. If the peg can only bear one half of the sum of the weights of m_1 and m_2, show that the least ratio of m_1 to m_2, consistent with the conditions of the system, is $3 + 2\sqrt{2}$.

(2.) A weight P, having fallen through a certain height, begins to pull up a heavier weight Q by means of a chord passing over a pulley; find the height through which it will lift it.

(3.) The time of descent of a weight of 12 lbs. down a plane, inclined at 30° to the horizon, is doubled by its connection with a weight hanging by a string passed over a pulley at the top of the plane; what is the latter weight?

(4.) Two particles of given masses are connected by an inextensible string, which is laid over a double inclined plane with a pulley at the top; and the planes are inclined to the horizon at angles a and β. Find the acceleration of the particles, and the tension of the string at any time.

(5.) Sixteen balls of equal weight are strung like beads upon a string, some of them are placed on an inclined plane, whose angle is $\sin^{-1}\frac{1}{3}$, and the rest hang over the top of the plane. How have the balls been arranged, if the acceleration of the resulting motion at first be half that of gravitation?

(6.) Sixteen equal weights are strung loosely on a string; how must they be arranged so that, when the string is laid upon a smooth fixed pulley, the motion may be the same as that produced when half the number of the balls is drawn over a smooth horizontal table by the weight of the other half hanging over the table edge.

(7.) A string hangs over a fixed pulley; a weight of two pounds hangs at one end, and a pulley at the other: over the pulley hangs a string carrying a weight of one pound at each end; when the whole is in equilibrium, any force is applied to one of the smaller weights; show that when it has pulled it down three inches, the other one pound weight and the two pound weight has each risen one inch; show also that, if the motion of the weight to which the force was applied be stopped in any gradual manner, the whole will be brought to rest, and the distances travelled by the weights will be as $3 : 1 : 1$.

(8.) If two equal masses be hanging, one at each end of a **string passing** over a smooth fixed pulley, and one be projected upwards with a velocity of $\frac{g}{4}$ feet per second, find when **the** string will become stretched, and the common velocity at the instant after **it** becomes so.

(9.) A number of equal weights are attached to different points of a string, and the string is then placed over a smooth pulley; show that, at any subsequent time, the tensions of **the** successive portions of the **string are, on each** side of the pulley, in arithmetic progression.

(10.) A string passes over a smooth fixed pulley carrying a weight P at one end and a pulley of weight Q at the other. Over the pulley Q is hung **a** string carrying weights p and q at its two ends respectively. Supposing P to move downwards, determine the tension **of** the two strings and the acceleration **of each body.**

(11.) A string, loaded with a series **of** equal weights at equal **distances** along it, is coiled up in the hand and held close to the **peg,** to which one end of the string is attached. The support of the **hand** being suddenly withdrawn from the coil, find the *finite* and *impulsive* strains on the peg **when the** r^{th} section **of** the string becomes **tight**; the mass of the string being neglected.

If a uniform chain, of length a and weight w, be treated **in a** similar manner, show that the strain **on** the peg when a length x of chain becomes tight $= 3 \cdot \dfrac{x}{a} W$.

(12.) Two weights are attached to the ends of an inextensible string, which **is hung over a** smooth pulley, and are observed to move through 6·4 feet in one second; **the** motion is then stopped, and a weight of five pounds attached **to** the smaller weight, **when** these descend through the same **space as** it ascended before in the same time. Determine the original **weights.**

(13.) Two weights **of five** pounds and **four** pounds together **pull** one of seven pounds **over** a smooth fixed **pulley,** by means of **a** connecting string; and **after** descending through a given **space** the four **pound** weight is detached and deposited without interrupting the motion. Through what **space** will **the** remaining **five pounds** descend?

(14.) Two scale-pans of equal weight W are connected by a fine string which passes over a smooth small pulley, and in them are placed weights W_1, W_2; show that the pressures, which these weights produce on the pans during motion, are

$$2W_1 \frac{W_2 + W}{W_1 + W_2 + 2W}, \text{ and } 2W_2 \frac{W_1 + W}{W_1 + W_2 + 2W}, \text{ respectively.}$$

(15.) A smooth uniform string hangs at rest over a peg. From one end of the string one fourth of its whole length is cut off. Show that the pressure on the peg is instantaneously diminished by one third of the whole weight of the string.

(16.) A fine string is attached to a fixed point, carries a small ring whose weight is W, and, passing over a small pulley in the same horizontal plane as the fixed point, has a weight $W_1 + W_2$ attached. The system being in equilibrium, the weight W_2 is removed. Show that the strain on the fixed point is instantly reduced by $\dfrac{W_2(W_1 + W_2)}{(W_1 + W_2)^2 + W_1 W}$ times its former value.

(17.) If a body be projected down a plane, inclined at an angle $30°$ to the horizon, with a velocity $= \frac{3}{4}$ of that due to the height of the plane, the time down the plane will equal the time down its vertical height from rest.

(18.) A particle falls down a smooth inclined plane. At the first observation the velocity is 25 feet per second, at another, three seconds later, it is 45; determine the inclination of the plane.

(19.) A body, moving down a smooth inclined plane, is observed to fall through equal spaces, a, in consecutive intervals of time T_1, T_2; prove that the inclination of the plane to the horizon is $\sin^{-1}\left(\dfrac{2a}{gT_1 T_2} \dfrac{T_1 - T_2}{T_1 + T_2}\right)$.

(20.) On a railway where the friction is $\frac{1}{240}$th of the load, show that five times as much can be carried on the level as up an incline of 1 in 60 by the same power at the same rate.

(21.) AP, AQ are two inclined planes, of which AP is rough ($\mu = \tan PAQ$) and AQ is smooth, AP lying above AQ; show that if bodies descend from rest at P and Q they will arrive at A, (1) in the same time if PQ be perpendicular to AQ, (2) with the same velocity if PQ be perpendicular to AP.

(22.) Two equal weights are connected by a string and laid on a table so that one is just over the edge and the other on the table, the string being stretched between them, perpendicular to the edge. If the length of the string be l and the height of the table h, find the velocity with which the second weight leaves the table.

(23.) A railway carriage detached from a train going at the rate of 30 miles an hour is stopped by the friction of the rails in half a minute; find the coefficient of friction.

(24.) A weight P after falling freely through a feet begins to raise a weight Q greater than itself, and connected with it by means of a string passing over a smooth fixed pulley. Show that Q will have returned to its original position after an interval

$$\frac{2P}{Q-P}\sqrt{\frac{2a}{g}}.$$

(25.) The time down a chord, to the vertex, of a parabola, whose axis is vertical, varies as the cosecant of the chord's inclination to the vertical.

(26.) Prove that the locus of the points, from which the times down equally rough inclined planes to a fixed point vary as the lengths of the planes, is a right circular cone.

(27.) A chord AB of a circle is vertical and subtends at the centre an angle $2\cot^{-1}\mu$. Show that the time down any chord AC drawn in the smaller of the two segments, into which AB divides the circle, is constant, AC being rough and μ the coefficient of friction.

(28.) AB is a vertical diameter of a circle, AP a chord meeting, when produced, the tangent at B in the point Q; prove that the time down $PQ \propto BQ$, and that the velocity acquired down $PQ \propto$ the chord BP.

(29.) Find the position of a point in the circumference of a circle, in order that the time of descent from it to the centre may be the same as the time of descent to the lowest point.

(30.) The plane of a circle is gradually inclined to the vertical. Show how the radius of the circle must change so that the time of descent down the chords may be the same as it was when the plane was vertical.

(31.) Determine the lines of quickest descent in the following cases :—

(1) From a point to a straight line ;
(2) From a circle to an external point in its plane ;
(3) To a circle from an external straight line in its plane.

(32.) A parabola is placed with axis vertical and vertex upwards ; prove that the square of the time of quickest descent from a given point in the axis along a chord to the curve varies as the sum of the latus rectum and the horizontal chord through that point.

(33.) Prove that, if PQ be a chord of quickest descent from one curve in a vertical plane to another, the tangents at P and Q are parallel, and PQ bisects the angles between the normals and the vertical.

(34.) Show that the time of quickest descent from any point of an ellipse to the horizontal axis major down the normal is $\sqrt{\dfrac{2le}{g}}$, l being the latus rectum, and e the eccentricity.

(35.) If two parabolas be placed with their axis vertical, vertices downwards and foci coincident, prove that there are three chords down which the time of descent of a particle, under the action of gravity, from one curve to the other, is a minimum, and that one of those is the principal diameter and the other two make an angle of 60° with it on either side.

(36.) The time of descent from rest down chords of an ellipse through the lowest point is a maximum, and a minimum, for those chords which are parallel to the transverse, and conjugate, axis, respectively.

(37.) In an Atwood's machine, if the string can only bear a strain of one fourth the sum of the weights at its two ends, show that the larger weight cannot be much less than six times the smaller, and that the least acceleration possible is $\dfrac{g}{\sqrt{2}}$.

(38). Compare the weights in an Atwood's machine when the heavier, starting from rest, descends a space equal to the length of the seconds pendulum (39·2 in.) in a second.

(39.) The power and weight are in equilibrium in the system of pulleys in which each hangs by a separate string. If the weight be doubled, prove that the acceleration of the weight is $\dfrac{g}{2^n + 2}$, the pulleys being without weight.

In the same system, the power and the weight (W) would be in equilibrium, if the pulleys were without weight; show that, if the weight of each pulley be w, the acceleration of W

$$= g\frac{3w}{3\,W.(2^n + 1) + w(2^{2n} - 1)}.$$

(40.) If the weight attached to the free end of the string in a system of pulleys, in which the same string passes round each of the pulleys, be m times that which is necessary to maintain equilibrium, show that the acceleration of the ascending weight is $\dfrac{m-1}{mn+1}g$, where n is the number of strings at the lower block, and the grooves of the pulleys are supposed perfectly smooth. Compare the tension of the string with the ascending weight.

(41.) Two particles slide down two straight lines, in a vertical plane, starting simultaneously from their point of intersection; prove that the line joining them at any time is equal to the space, through which a particle would have moved in the same time, along a line, whose inclination to the horizon is the angle between given lines.

(42.) In the first system of pulleys there are four moveable pulleys, each weighing one pound. Show that, if the power is given by a man hanging on with a uniform strain of 150 lbs., a ton would be raised from rest 903·6 feet in the first half minute.

(43.) Two unequal weights are connected by an inextensible string which passes over a smooth fixed pulley. The stand supporting the pulley is placed in the scale-pan of a balance. Show that the system will have a constant apparent weight during the motion, and compare it with the true weight.

146. PROP. I. *When a heavy particle is projected in any direction, not vertical, its path is a parabola.*

The only force acting on the particle is its weight, and this, by Law II., will produce the same change in the motion of the particle as if it were initially at rest.

Suppose then the particle to start from a point P, in direction PT, with a velocity u. After a time t it would, if initially at rest, have *acquired* a vertical velocity gt; and therefore in the given case it must also have *acquired* a vertical velocity gt. Hence its velocity can be represented by saying that it has one velocity u in direction PT and another gt vertically downwards; and the same is the case whatever value t have.

Therefore, it always moves in the vertical plane through PT; and we can represent its motion (Art. 50) by saying that it retains its velocity u parallel to PT and has the acceleration g vertically downwards.

Draw MPV vertically through P.

Let Q be the position of the particle after any time t reckoned from the instant of starting.

Draw QV parallel to PT, meeting MPV in V,
and $\quad Qm \quad$,, $\quad MPV \quad$,, $\quad PT \quad$,, m.

Then, Art. 62, (1), $\quad PV = Qm = \frac{1}{2}gt^2$,

$$VQ = Pm = u.t.$$

$\therefore$, eliminating t, $QV^2 = \dfrac{2u^2}{g} \cdot PV.$ . . . (1)

Also, the particle begins by moving along PT, *i.e.* PT is the tangent to the path at P.

Now, if we drew a parabola, passing through P, having PT as the tangent and PV as the diameter at P, and its focus S at

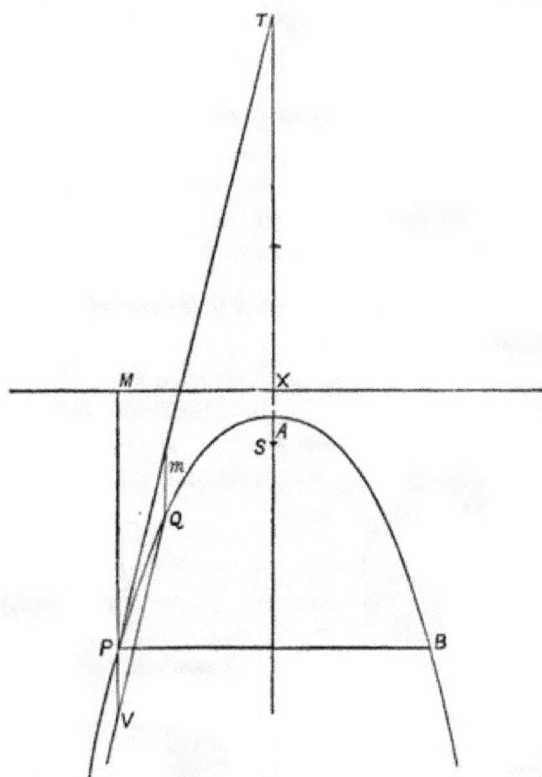

FIG. 39.

a distance $\dfrac{u^2}{2g}$ from P, then (1) is just the relation which would hold between the ordinate QV and the abscissa PV of any point Q on it. Hence we conclude that the path is this parabola.

Cor. Let VPM meet the directrix in M. Then $PM = PS$

$$= \frac{u^2}{2g}.$$

147. Prop. II. *The velocity at any point of the path is the same in magnitude as would be acquired by a particle in falling to that point from the directrix.*

Let V be the velocity acquired by a particle in falling from M to P. Then $V^2 = 2.g.PM = 2.g.\dfrac{u^2}{2g} = u^2$.

Hence the Proposition is true for the point of starting.

Now the particle must start from every point of its path in order to describe the subsequent portion. Therefore every point may be considered as a point of starting.

Hence the Proposition is true for all points in the path.

148. Let θ be the angle which PT makes with the horizon, called the angle of projection.

Then u, the velocity of projection, is equivalent to

      (1) $u \cos\theta$ in the horizontal direction, and

      (2) $u \sin\theta$,,   vertical   ,,   .

By discussing separately the motions in these two directions, we can work out most problems.

149. Consider what takes place during the motion. The force which acts on the particle changes the velocity in the vertical direction only.

Hence the horizontal component remains constant and equal to $u \cos\theta$.

Also, the effect of the force is to generate during every second a change, equal to g vertically downwards, in the velocity. Hence the vertical component at the end of the time t is $u \sin\theta - gt$.

If the particle is projected upwards, $u \sin\theta$ is positive; and therefore $u \sin\theta - gt$ is positive at first; but as t increases it gradually diminishes till it vanishes; after that it becomes negative, showing that the particle is now descending.

When $u \sin\theta - gt = 0$, the particle is at its highest point (A). At this instant, there being no vertical velocity, it is moving horizontally, *i.e.* the tangent to the path at A is at right angles to the axis, and therefore A is the vertex of the parabola.

150. Prop. III. *To find the time of reaching the vertex, and its height, when there, above the point of starting.*

Let T be the time of rising to A, then $u \sin\theta - g.T. = 0$;

$$\therefore T = \frac{u \sin\theta}{g}.$$

Let h be the vertical height of A. On starting from P the particle has a vertical velocity $u \sin\theta$, and, during the interval that it is traversing the vertical distance h, it has a vertical acceleration $-g$, and at the end its vertical velocity is 0;

$$\therefore \quad 0 = u^2 \sin^2\theta - 2gh, \text{ Art. } 58 \text{ (3)};$$

$$\therefore h = \frac{u^2 \sin^2\theta}{2g}.$$

151. Prop. IV. *To find the latus rectum.*

Let the axis SAX meet the directrix in X, then the whole latus rectum $= 4AX$.

At A the particle has no vertical velocity; therefore its whole **velocity is the constant** horizontal velocity $u \cos\theta$; also, **its distance from** the directrix is AX.

Now the velocity acquired in falling from A to $X = \sqrt{2g AX}$;

$$\therefore \quad u^2 \cos^2\theta = 2.g.AX; \text{ (Art. 147.)}$$

$$\therefore \quad 4AX = \frac{2u^2 \cos^2\theta}{g}.$$

EXAMPLES.—XXIII.

(1.) A fine tube inclined to the vertical **has a** uniform rectilinear motion parallel to **itself,** and a particle is allowed **to** run down it; find **the** locus of the path of the particle in space.

(2.) A particle is **projected with** a velocity, 60 in a direction making 60° with the horizon. Determine the **position** of the directrix and the length of the latus rectum of **its path.**

(3.) A ball is projected **in** a direction inclined at an angle of 30° to the horizon, and with a velocity which it would have acquired in falling from rest through a space of 100 yards; find **the** greatest height attained by the ball.

(4.) A body projected from A is describing a parabola, and when it is at P it is vertically above B, which is a point in the horizontal plane through A. If the direction of motion at P cuts the line of projection in the point Q, prove that $QA = QB$.

(5.) A number of heavy particles are thrown all from one point, so as all to attain the same greatest height in the same vertical plane. Show that the locus of the foci of the curves described is a parabola whose vertex is at the same height above the point of projection.

(6.) A number of particles are projected from the same point, so as all to describe parabolas having the same latus rectum. Show that the locus of the foci is an equal parabola, with its vertex downwards, and its focus at the point of projection.

(7.) In the path of a projectile, if u and v be the velocities at the ends of any focal chord, and V_x the horizontal velocity of projection, show that $\dfrac{1}{u^2} + \dfrac{1}{v^2} = \dfrac{1}{V_x^2}$.

152. Prop. V. *To determine the motion at any instant.*

Let t be the time from the moment of starting till the instant under consideration.

Let v be the velocity at this instant,

ϕ the acute angle the direction of motion then makes with the horizon.

Then $v \cos\phi =$ the horizontal component of the velocity
$$= u \cos\theta.$$

And $v \sin\phi =$ the vertical component of the velocity
$$= u \sin\theta - gt.$$

Therefore $\tan\phi = \dfrac{u \sin\theta - g.t.}{u \cos\theta}$.   .   .   .   (1);

and $v^2 = u^2 - 2g.u \sin\theta.t + g^2 t^2$.   .   .   (2).

Thus (1) and (2) give the direction and magnitude of the velocity at the given instant.

153. Prop. VI. *To find the range, and time of flight, on the horizontal plane through the point of starting.*

From P draw PB horizontally to meet the curve in B (Fig. 39).

Let t be the time from P to B, and $R = PB$.

Then we have to find t and R.

Now the particle starts from P with the vertical velocity $u \sin\theta$, and during the time t it has a vertical acceleration $-g$, and at the end of t its vertical distance from P is 0;

$$\therefore 0 = u \sin\theta.t - \tfrac{1}{2}gt^2, \text{ Art. 58 (2)};$$

$$\therefore t = \frac{2u \sin\theta}{g}.$$

Also, it starts from P with the horizontal velocity $u \cos\theta$, which it retains throughout the time t, and at end of t its horizontal distance from P is R;

$$\therefore R = u \cos\theta.t = \frac{2u^2 \sin\theta \cos\theta}{g} = \frac{u^2 \sin2\theta}{g}.$$

Cor. 1. In Art. **150** we found $T = \dfrac{u \sin\theta}{g}$; $\therefore t = 2T$.

Therefore the time of going from P to B is double the time from P to A, and therefore the time from A to B is equal to the time from P to A.

Cor. 2. Also, its distance from the directrix is the same at B as at P; therefore (Art. **147**) its velocity must be the same in magnitude at B as at P. And by the symmetry of the parabola its directions of motion, i.e. the tangents at B and P, must make the same acute angle with PB.

154. PROP. VII. *To find the range, and time of flight, on any plane through the point of starting.*

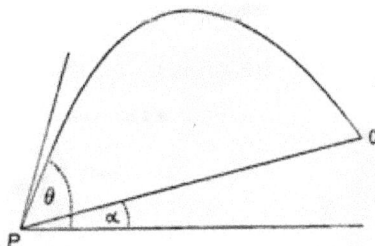

FIG. 40.

Let the plane meet the **vertical** plane of the path in the line PC, and let PC meet the path again in C.

Let t be the time of going from P to C, and $R=PC$. We have to find t and R.

We will consider separately the motion in two directions at right angles, viz., parallel and perpendicular to PC.

Let a be the inclination of PC to the horizon.

The velocity of starting is equivalent to

$$u \cos(\theta-a), \text{ parallel to } PC,$$
$$\text{and } u \sin(\theta-a), \text{ perpendicular to } PC.$$

The acceleration throughout the motion is equivalent to

$$-g \sin a, \text{ parallel to } PC,$$
$$\text{and } -g \cos a, \text{ perpendicular to } PC.$$

At the end of the time t, the distances from P are

$$R, \text{ parallel to } PC,$$
$$\text{and } 0, \text{ perpendicular to } PC.$$

Therefore $R=u \cos(\theta-a)t-\tfrac{1}{2}g \sin a t^2$ (1),

and $0=u \sin(\theta-a)t-\tfrac{1}{2}g \cos a t^2$ (2).

From (2) $t=\dfrac{2u \sin(\theta-a)}{g \cos a}$.

Therefore, from (1),

$$R=\frac{2u^2 \cos(\theta-a) \sin(\theta-a)}{g \cos a} - \frac{2u^2 \sin^2(\theta-a)}{g \cos^2 a} \sin a$$

$$=\frac{2u^2 \sin(\theta-a)}{g \cos^2 a}\left\{ \cos(\theta-a) \cos a - \sin(\theta-a) \sin a \right\}$$

$$=\frac{2u^2 \sin(\theta-a) \cos\theta}{g \cos^2 a}.$$

COR. Putting R into the equivalent form

$$\frac{u^2}{g \cos^2 a}\left\{ \sin(2\theta-a)-\sin a \right\},$$

we see that R is greatest when $\sin(2\theta-a)$ is greatest,

i.e. when $\sin(2\theta-a)=1$,

i.e. when $2\theta-a=\dfrac{\pi}{2}$,   i.e. when $\theta=\dfrac{\pi}{4}+\dfrac{a}{2}$.

Thus the greatest range is $\dfrac{u^2}{g(1+\sin a)}$.

(1.) A particle is projected with a velocity g at an inclination of 30° to the horizon. Find the magnitude and the direction of the velocity at the end of 10″.

(2.) A body is projected in a direction inclined to the horizon at an angle of 15°, with a velocity 20. Determine the range and time of flight on the horizontal plane.

(3.) A body starts with a velocity $3g$ at an angle 75° to the horizon. Find the times when it will be 30 feet above the point of starting, and the distance between its positions at those times.

(4.) A particle is projected with the velocity 100 at an angle 45° to the horizon. Find its range and time of flight on a plane through the point of starting inclined at an angle 30° to the horizon.

(5.) A particle is started with a velocity 20. Find its greatest possible range on the horizontal plane through the starting point.

(6.) Find the greatest range of a projectile on an inclined plane through the point of projection; the initial velocity being 21, and the inclination of the plane 30°.

(7.) From the top of a hill, inclined to the horizon at an angle 30°, a ball is projected with a velocity v at an acute angle to the hill. Find the greatest range down the hill.

(8.) If a body be projected at an angle a to the horizon with the velocity due to gravity in 1″, its direction is inclined at an angle $\frac{a}{2}$ to the horizon at the time $\tan\frac{a}{2}$, and at an angle $\frac{\pi-a}{2}$ at the time $\cot\frac{a}{2}$.

(9.) With what velocity must a projectile be fired at an elevation of 30° so as to strike an object at the distance of 2600 feet in an ascent of 1 in 39?

(10.) The greatest range of a rifle on level ground is 1176·3 feet. Find the initial velocity of the ball, and show that the greatest range up an incline of 30° will be 784·2 feet, neglecting the resistance of the atmosphere.

(11.) The greatest range of a rifle ball up an incline of 30° is found to be 3921 feet; find the initial velocity of the ball, and show that its greatest range on level ground would have been 5881·5 feet, neglecting the atmospheric resistance.

(12.) A particle is projected from the top of a tower with the velocity which would be acquired in falling vertically down n times the height of the tower; find the range on the horizontal plane through the bottom of the tower, and show that it will be a maximum when the angle of projection is $\frac{1}{2} \sec^{-1}(1 + 2n)$.

(13.) Prove that the least angle of inclination to the horizon, at which a particle can be projected so as to strike at right angles any plane through the point of projection, is $\cos^{-1}\frac{1}{3}$.

(14.) Two projectiles fired with velocities due to the heights h_1, h_2, at angles of elevation from the horizon e_1, e_2, strike the same point on the side of a hill on which the gun is placed; find the inclination of the hill to the horizon.

(15.) A shell fired at an elevation θ from a mortar placed at A, just clears a vertical wall whose elevation at A is a, and strikes the ground beyond at B; show that the horizontal range AB is divided by the wall in D so that $AB : BD = \tan\theta : \tan a$.

(16.) Two bodies, projected from the same point A, in directions making angles a, a' with the vertical, pass through the point B in the horizontal plane through A; prove that, if t, t' be the time of flight from A to B,

$$\frac{\sin(a - a')}{\sin(a + a')} = \frac{t'^2 - t^2}{t'^2 + t^2}.$$

(17.) If v be the velocity with which a particle is projected, t, t' the times it takes to reach the ends of a focal chord, ϕ the angle which its direction of motion at the time t makes with the horizon, then

$$t^2 \cos^2\phi + t'^2 \sin^2\phi = \frac{v^2}{g^2}.$$

(18.) Three particles are projected simultaneously from the same point, and strike the horizontal plane through that point simultaneously; prove that, if their ranges are in geometrical progression, the latera recta of their paths will also be in geometrical progression.

155. *Ex.* A particle is projected from a point P with a velocity of given magnitude. Find the direction of projection in order that it may pass through another given point C.

Let u denote the given velocity.

Draw PM vertically upwards from P, and equal to $\frac{u^2}{2g}$, and draw MN horizontally.

Then MN (Art. 146, Cor.) is the directrix of the necessary path.

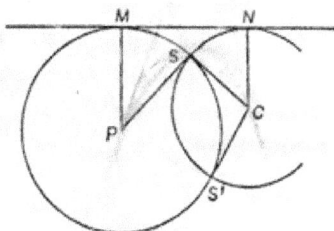

Fig. 41.

Draw CN perpendicular to MN, and describe two circles, with centres P and C and radii PM, CN, cutting each other in S and S'.

Then a parabola, described with focus S and directrix MN, and passing through P, will also pass through C, since the distance of C from the directrix is equal to its distance from the focus; and the tangent at P to this parabola is one of the directions along which the particle must start in order to reach C. But this tangent bisects the angle SPM; therefore this bisecting line is one of the necessary directions of projection.

Similarly we can show that the bisector of the angle $S'PM$ is another such direction.

If the circles touch, S and S' coincide in the point of contact, and there is only one direction of projection along which the particle can start from P so as to reach C.

If the circles do not meet there is no such direction.

156. The problem, which we have discussed in this chapter, is sometimes enunciated in its kinematical form as follows :—

A point is projected with a given velocity, and its motion has an acceleration constant in magnitude and direction, the latter not being coincident with that of the velocity of projection; determine the motion and the path.

I

The **work is** exactly the same. We should, however, use some general symbol (f) **for** the acceleration instead of g, and instead of "the vertical line," we should **talk** of "the constant direction of acceleration;" and, instead of a "horizontal line," "a line perpendicular to the constant direction of acceleration;" and instead of "falling from M to P," "moving from M to P with the constant acceleration."

157. The student is advised, in working the examples, **not** to assume any of the results proved in this chapter, **except** those of Art. **146.**

All other results should be worked out for each particular case.

158. *Ex.* **A** particle is projected on a smooth inclined plane, in some direction not in the line of greatest slope through the point of projection; determine the motion.

We may reproduce Art. 136 as far as (3); except that instead of PA we will draw PV, and instead of the phrase "down PA," we will use "parallel to PV."

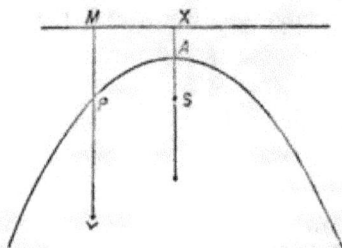

Fig. 42.

Further, the direction of projection is not coincident with the direction of the constant acceleration $g \sin\theta$.

Hence the path is a parabola on the inclined plane, having its axis in the direction of the lines of greatest slope. And all the results of Art. 146, etc. follow, reading $g \sin\theta$ for g, "line of greatest slope" for vertical line, etc., and "time of sliding from M to P" instead of "time of falling from M to P."

159. The most common instances, in which a body is projected in a direction not vertical, occur in the use of firearms.

There is a great difference between the lengths of the ranges found by the methods of this chapter and by measurement in practice. This difference is due to the resistance of the air, which has considerable effect in retarding the balls, and thus there is produced a great difference between the motions of projectiles in vacuo and in air.

The analysis, which the student is supposed in this treatise to have at his command, is not sufficient for calculating the effect of this resisting force of air.

EXAMPLES.—XXV.

(1.) *ABC* is a right-angled triangle in a vertical plane with its hypothenuse *AB* horizontal; a particle projected from *A* passes through *C* and falls at *B*; prove that the tangent of the angle of projection $= 2 \operatorname{cosec} 2A$, and that the latus rectum of the path described is equal to the height of the triangle.

(2.) A particle is projected, with a velocity $\sqrt{2ga}$, from a point *A* so as to pass through another point *B*, whose vertical and horizontal distances from *A* are h, k; show that, when there is only one possible direction of projection, $k^2 = 4a(a - h)$.

(3.) Two inclined planes of equal altitude h, and inclined at the same angle a to the horizon, are placed back to back upon a horizontal plane. A ball is projected from the foot of one plane along its surface and in a direction making an angle β with its line of intersection with the horizontal plane. After flying over the top of the ridge it falls at the foot of the other plane; show that the velocity of projection is

$$\tfrac{1}{2}\sqrt{gh} \, \operatorname{cosec}\beta\sqrt{8 + \operatorname{cosec}^2 a}.$$

(4.) If any number of particles be projected at the same time and in the same direction with different velocities, show that at any time a straight line can be drawn through them; and, if in different directions with the same velocity, that they all lie at any time on the surface of the same sphere.

(5.) A particle is projected up a smooth inclined plane in the form of a rectangle with given sides. Find the velocity of projection from one corner in order that the particle may leave the plane horizontally at the other corner; and show that the ratio of the horizontal range after leaving the plane to that described on the plane is the sine of the angle of elevation of the plane.

(6.) A shell explodes at its highest point into two equal parts which receive equal velocities in opposite directions, supposing that the times which elapse between the explosion and the arrival of each piece on the ground be observed, and also the distances of the pieces from the original point of projection, find the velocities and directions of explosion and projection.

If the times be as the squares of the distances, the direction of explosion will be vertical.

(7.) Particles are projected from the same point and in the same vertical plane so as to describe equal parabolas. Show that the vertices of their paths lie on a parabola.

(8.) Chords are drawn, joining any point of a vertical circle with its highest and lowest points; prove that, if a heavy particle slide down the latter chord, the parabola, which it will describe after leaving the chord, will be touched by the former chord, and that the locus of the points of contact will be a circle.

(9.) A body is projected up an inclined plane, whose inclination (a) to the horizon is less than $45°$, with a velocity due to a height h. Find the length of the inclined plane in order that the distance, at which it strikes the horizontal plane through the point of starting, may be a maximum. Prove that the length is $2h \cot 2a \sec a$.

(10.) A heavy particle is projected from one point so as to pass through another, not in the same horizontal line with it; prove that the locus of the focus of its path will be an hyperbola.

(11.) If tangents be drawn to two parabolic paths, having the same foci, from any point in the common axis, the velocities at the points of contact are equal.

(12.) A cat on the top of a wall of height h springs at a mouse on a horizontal plane below at a distance n from the wall; the mouse runs at once towards its hole which is vertically under the cat's position. The cat just catches the mouse; show that the velocity of the mouse

$$= \tfrac{1}{2} \sqrt{\frac{g}{2h}} \left\{ \sqrt{4n^2 + hl} - \sqrt{hl} \right\},$$ where l is the latus rectum of the cat's path.

(13.) Two bodies are projected from the same point at the same instant, with velocities v_1 and v_2, and in directions making angles a_1 and a_2 with the horizon ; show that the time, which elapses between their transits through the other point which is common to their paths,

$$= \frac{2}{g} \cdot \frac{v_1 v_2 \sin(a_1 \sim a_2)}{v_1 \cos a_1 + v_2 \cos a_2}.$$

(14.) Show that the product of the velocities at any two points of the path of a projectile is proportional to the distance of the point of intersection of tangents at those points from the focus.

(15.) A body is projected with a velocity V in a direction OP inclined at an angle a to the horizon, and a horizontal velocity $V \sin a$ in a direction at right angles to OP. Show that the latus rectum of the path described is equal to 4 times the greatest height the body would rise to, if projected vertically upwards with the velocity V.

(16.) A stone is thrown in such a manner that it would just hit a bird at the top of a tree, and afterwards reach a height double that of the tree. If at the moment of throwing the stone the bird flies away horizontally, prove that the stone will, notwithstanding, hit the bird, if its horizontal velocity be to that of the bird as $\sqrt{2} + 1 : 2$.

(17.) Two particles are projected from the same point on an inclined plane, one horizontally and the other at right angles to the horizontal lines, find their distance apart at any time.

(18.) Show that the two instants, at which a body has a certain angular elevation when seen from one point in the plane of its motion, are equidistant from the two instants at which it has the same angular elevation when seen from another point in the same plane on a level with the former.

(19.) A particle slides down a smooth inclined plane ; prove that the distance between the foot of the plane and the focus of the particle's path after leaving the plane is equal to the height of the plane.

(20.) A rifle sighted to hit the centre of the target at a distance of a, and on the same horizontal line as the muzzle, is inclined at an angle a to the horizon. Show that if its inclination to the horizon receive a small decrease θ, the ball will hit the target at a point $\dfrac{2a \cos 2a}{1 + \cos 2a}\theta$ above the centre.

(21.) A heavy particle is projected obliquely upwards ; if v_1, v_2 be its component vertical velocities ascending, and v_3, v_4 descending, respectively, at any four points of its path which lie on a circle, show that $v_1 + v_2 = v_3 + v_4$.

160. In this chapter we shall discuss the change produced in the motion of a body by its impinging, or colliding, against another, or by another impinging, or colliding, against it. The act of impinging, or colliding, is called impact, or collision.

161. When two bodies come into collision, they may or they may not afterwards separate. At any rate a finite change is almost instantaneously produced in the motion of each, and therefore each must have exerted on the other an impulsive force to produce this change. The two forces thus brought into play are, by Law III., equal and opposite.

Let the bodies be smooth spheres, A and A'. Then, since they are smooth, the mutual action between them must be wholly normal, and, since they are spheres, it must be in the line joining their centres, called their line of centres.

Then, by Law II., the only change, which takes place in the motion of either, must be in this line.

162. *Direct Impact.*—In the first place let their centres be moving before impact in this straight line; and therefore after impact the centres will continue to move in it.

Let m and m' denote the masses of A and A'. Suppose that they are both moving one way before impact, say from left to right.

Let u and u' denote their velocities before impact.

[The case of one ball *meeting* the other would be allowed for by changing the sign of one velocity, say u'.]

Let v and v' denote the velocities **after** impact.

[Thus v will be positive, or negative, according as A is moving after impact **from left to right**, or from right to left. Similarly for v' and A'.]

Then $mu-mv$ denotes the change in the momentum of A,

$m'u'-m'v'$ „   „   „ A'.

Let R denote the amount of force exerted by A on A', and therefore $-R$ the amount exerted by A' on A. These produce the above changes.

Therefore, Art. **123**, $mu-mv=-R$.   .   .   (1),

$$m'u'-m'v'=R \quad . \quad . \quad . \quad (2).$$

These equations are not sufficient to determine v, v' and R. To find a third equation we must **consider** whether or not the bodies tend to separate after impact, *i.e.* whether the bodies are elastic or inelastic.

I. *Inelastic Bodies.*—No separation takes place **after** impact between such bodies. Hence after impact they both move on with the same velocity ;

$$\therefore v=v' \quad . \quad . \quad . \quad . \quad (3).$$

Now from (1), $u-v=-\dfrac{R}{m}$,

and from (2), $u'-v'=\dfrac{R}{m'}$;

subtracting, we have, since $v=v'$, $u-u'=-\dfrac{R}{m}-\dfrac{R}{m'}$;

$$\therefore R=-\frac{mm'}{m+m'}(u-u') ; \quad . \quad . \quad . \quad (4).$$

$\therefore$, from (1), $v=u+\dfrac{R}{m}=u-\dfrac{m'}{m+m'}(u-u')$

$$=\frac{mu+m'u'}{m+m'} \quad . \quad . \quad . \quad (5).$$

This last result could also have been obtained thus. Since, by Art. **128**, no change can be produced in the momentum of the whole system by the impact, therefore $mv+m'v'=mu+m'u'$.

II. *Elastic Bodies.*—It is found by experiment that, as long as the substances of which the bodies are made remain the same, whatever be their masses and their velocities before impact, their relative velocity after impact bears *a constant ratio* to their relative velocity before impact. This constant ratio is called the *modulus of elasticity* of the two substances. It is usually denoted by e.

Thus, when iron impinges on iron, e has one value,

    ,,   iron   ,,   wood, e has another value,

    ,,   wood ,,   wood, e has a third   ,,

The different values of e can be determined by experiments which will be described in a future chapter. In our case, before impact, since both balls are supposed to be moving the same way and A is catching up A', $u-u'$ denotes the relative velocity. For the positive case after impact we should suppose that A and A' both move the same way as before impact, and then $v'-v$ will represent their relative velocity.

[If A were to move the opposite way after impact, v would be negative, and the relative velocity would be the sum of their actual velocities.]

Hence $v'-v : u-u' = e$; $\therefore v'-v = e(u-u')$.   .   (6).

Now, from (1), $u-v = -\dfrac{R}{m}$,

and from (2), $u'-v' = \dfrac{R}{m'}$;

$\therefore$, subtracting, $-\dfrac{R}{m} - \dfrac{R}{m'} = u-u'-v+v' = u-u'+(v'-v)$

$$= u-u'+e(u-u') = (1+e)(u-u');$$

$$\therefore R = -\frac{mm'}{m+m'}(1+e)(u-u'); \quad . \quad (7).$$

$\therefore$, from (1), $v = u + \dfrac{R}{m} = u - \dfrac{m'}{m+m'}(1+e)(u-u'),$.   (8),

from (2), $v' = u' - \dfrac{R}{m'} = u' + \dfrac{m}{m+m'}(1+e)(u-u')$.   (9).

163. If for any two substances e were equal to 1, they would be said to be perfectly elastic. Probably no such substances exist in nature.

164. If we put $e=0$, the formulæ for elastic bodies become the same as those for inelastic bodies. This we should expect beforehand, for when the bodies are inelastic $v'=v$, or $v'-v=0$, and therefore $e=\dfrac{v'-v}{u-u'}=0$.

165. When two bodies impinge as above, so that at the moment of impact neither has any motion except in the direction of the common normal, the impact is said to be *direct*. In other cases the impact is said to be *oblique*.

EXAMPLES.—XXVI.

(1.) A sphere of six pounds mass, moving at the rate of 10 miles an hour, overtakes another of four pounds mass, moving at 5 miles an hour; determine their velocities after collision, assuming $e=\frac{1}{2}$, the impact being supposed direct.

(2.) If A impinges on B at rest, and is itself reduced to rest by the impulse, find the ratio of the masses of A and B, when the elasticity $=\frac{1}{3}$.

(3.) The result of a direct impact between two balls moving with equal velocity is such that one of them returns with its former velocity, and the other follows it with half that velocity. Show that one ball is four times as heavy as the other, and that $e=\frac{1}{4}$.

(4.) Two perfectly elastic balls, of masses m and $3m$, meet one another when travelling with velocities u and $5u$. Find their subsequent velocities.

(5.) A ball impinges on another, of twice its mass, travelling in the opposite direction with two-thirds of its velocity ($e=\frac{1}{2}$). Show that it returns with one-third of its former velocity.

(6.) A ball impinges on another at rest, of half its mass, which afterwards travels with a velocity 4. Find the original and subsequent velocities of the first ball, e being $\frac{1}{2}$.

(7.) Three balls, A, B, C are placed in a line, A, B are of the same mass. If A strikes B directly, show that after B has struck C, A will overtake B, if C's mass be more than $\dfrac{2e}{1+e^2}$ of A's, e being the common coefficient of elasticity.

(8.) A, B, C are three perfectly elastic balls at rest in the same straight line, B is made to impinge upon A and rebounding strikes C. Show that, if A and C, after having been struck by B, each move with the same velocity, $m_2 + m_3 = m^1 - m_2$, m_1, m_2, m_3 being the masses of A, B, C respectively.

(9.) A series of perfectly elastic balls are arranged in the same straight line, one of them impinges directly on the next, and so on ; prove that, if their masses form a geometrical progression of which the common ratio is 2, their velocities after impact will form a geometrical progression of which the common ratio is $\frac{2}{3}$.

(10.) Four perfectly elastic bodies, A, B, C, D, are situated in a straight line, the three last being at rest ; find the ratio of their masses so that the quantity of motion in A may be equally divided among the four balls after collision.

166. *Oblique Impact.*—Let two spheres, A and A', impinge obliquely.

Let the directions of motion of the two centres make, with the line of centres, angles a, a' before, and β, β' after, impact.

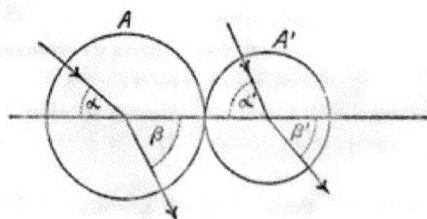

Fig. 43.

Then the velocity of A, before impact, is equivalent to $u \cos a$ in the line of centres and $u \sin a$ perpendicular to this line ;

and, after impact, to $v\cos\beta$ and $v\sin\beta$ in these directions. Similarly for the velocities of A'.

Now the only change, which takes place on account of the impact, being in the direction of the line of centres, the resolved part of the velocity, of either ball, perpendicular to this line will be the same after impact as before;

$$\therefore v\sin\beta = u\sin\alpha, \qquad \cdot \qquad \cdot \qquad (1),$$
$$v'\sin\beta' = u'\sin\alpha' \qquad \cdot \qquad \cdot \qquad (2).$$

Again, by Law II., the change of motion in the line of centres is the same as if the motion perpendicular to this line did not exist; *i.e.* it is the same as if the impact had been direct, and the spheres had been moving with velocities $u\cos\alpha$, $u'\cos\alpha'$.

Therefore, if the spheres are *inelastic*,

$$v'\cos\beta' = v\cos\beta = \frac{mu\cos\alpha + m'u'\cos\alpha'}{m+m'}. \qquad \cdot \qquad \cdot \qquad (3).$$

These, with (1) and (2), are sufficient for determining v, v', β, β'.

$$\text{Also, } R = -\frac{mm'}{m+m'}(u\cos\alpha - u'\cos\alpha') \qquad \cdot \qquad (4).$$

And, if the spheres are *elastic*,

$$v\cos\beta = u\cos\alpha - \frac{m'}{m+m'}(1+e)(u\cos\alpha - u'\cos\alpha'), \qquad (5),$$

$$v'\cos\beta' = u'\cos\alpha' + \frac{m}{m+m'}(1+e)(u\cos\alpha - u'\cos\alpha'). \qquad (6).$$

These, with (1) and (2), are sufficient for determining v, v', β, β'.

$$\text{Also, } R = -\frac{mm'}{m+m'}(1+e)(u\cos\alpha - u'\cos\alpha') \qquad (7).$$

EXAMPLES.—XXVII.

(1.) A sphere of mass 3 moving with a velocity 5 *overtakes* another sphere of mass 7 moving with a velocity 4. The directions of the first and second spheres before impact make angles 30° and 60° with the lines of centres at the instant of impact, and both spheres before impact are moving towards the line of centres from the *same* part. Determine their motions after impact, e being $\frac{1}{4}$.

(2.) A collision takes place between two balls of masses 5 and 6 *meeting* one another with velocities 3 and 4 in directions making angles 30° and 45° with the line of centres towards which they are moving before the collision from *opposite* parts. Determine their subsequent motions, the modulus of elasticity being $\frac{2}{3}$.

(3.) A ball moving with velocity 5 impinges at an angle 30° on a ball of double its mass at rest. The coefficient of elasticity being $\frac{1}{2}$, determine their subsequent motions, and the angle between their lines of motion.

(4.) One ball impinges on another at rest, of three times its own mass, the coefficient of elasticity being $\frac{1}{3}$; and after the impact the direction of its motion is found to be inclined at an angle of 60° to its old direction. Determine the angle between its original direction and the line of centres. If the other ball start with a velocity 3, determine the velocity of the first before and after impact.

(5.) A sphere impinges obliquely on another at rest; determine the angle of deviation through which the direction of its motion is turned.

(6.) Two balls, whose masses are m and m', impinge, and their directions of motion after impact are perpendicular to their directions before impact; if a, a' be the angles which their directions before impact make with the line joining their centres, prove that the elasticity is

$$\frac{m \sin^2 a' + m' \sin^2 a}{m \cos^2 a' + m' \cos^2 a}.$$

(7.) Two perfectly elastic balls, A and B, impinge upon each other. First A impinges on B at rest and goes off in a direction making an angle θ with the line joining their centres; then B impinges on A at rest and at the same angle of incidence and goes off at an angle θ. Prove that $\theta + \theta' = 180$.

Prove also that, if the balls are imperfectly elastic, and the angles of incidence in the two cases be a and a', then

$$\frac{\cot\theta}{\cot a} + \frac{\cot\theta'}{\cot a'} = 1 - e.$$

(8.) Two balls, of elasticity e, moving with equal momenta in parallel directions, impinge; prove that, if their directions of motion be opposite, they will move after impact with equal momenta in parallel directions, and that this direction will be perpendicular to the original direction, if their common normal is inclined at an angle $\sec^{-1}\sqrt{1+e}$ to that direction.

·(9.) Small equal spherical balls of perfect elasticity are placed at the angular points of a regular polygon of n sides; one of them is projected with velocity (V) so as to strike all the others in succession, and to pass through its original position. Find the velocity with which it returns, and the directions of motion and the velocities of all the balls.

(10.) A ball is struck so as to proceed at 45° to its previous direction of motion; what must have been its mass that the same blow might have diverted the direction at right angles?

(11.) $ABCD$ is an ordinary rectangular billiard table, perfectly smooth; E a ball in a given position; it is required to select the proper position for another ball F, in all respects like the first, so that the player striking E upon F, may cause F to run into the corner pocket A and E to run into D with equal velocities, without the intervention of the sides, the elasticity of the balls being perfect.

(12.) If a row of equal, perfectly elastic, balls in contact be struck directly by a similar ball, show that the last ball will fly off with a velocity equal to that of the striking ball, and that the others will remain at rest.

(13.) Two perfectly elastic balls, of equal size and mass, impinge, after describing given distances on a smooth horizontal plane with uniform velocities. Prove that, if the directions of motion after impact are parallel, the cosine of the angle between their original directions is equal to the ratio of the product of the velocities after and before impact. Show also that the rectangles, of which the described distances are diagonals, and whose sides are parallel and perpendicular to the line of impact, are equal in area.

167. *Impact of a sphere on a fixed surface.*

If an elastic ball impinge on a fixed smooth plane, we may deduce the subsequent motion of the ball from the preceding formulæ, as follows :—

Let m be the mass of the ball; u, v be its velocities before, and after, impact; a, β the acute angles which the directions of its motion, before and after impact, *make with the plane.*

Since the plane is fixed, its velocity before and after impact must be zero; $\therefore u'=0, v'=0$. Also it may be regarded as part of the earth and as a sphere of mass m' so large that $\dfrac{m}{m'}$ is so small that it may be neglected;

$$\therefore \frac{m'}{m+m'}=\frac{1}{\dfrac{m}{m'}+1}=1.$$

Thus $v \cos\beta = u \cos a$,

And $v \sin\beta = u \sin a - (1+e) u \sin a = - eu \sin a$;

$\therefore \tan\beta = e \tan a$, and $v = u\sqrt{e^2 \sin^2 a + \cos^2 a}$.

If the ball is inelastic, we have $e=0$ and $v\sin\beta=0$, or $\beta=0$, so that the ball after impact remains at rest when the impact is direct, and moves along the plane with a velocity $u \cos a$ when the impact is oblique.

If the ball be perfectly elastic, $e=1$; $\therefore a=\beta$; and $v=u$.

168. These results may also be obtained at once, as follows :—

The action of the plane on the ball is normal to the plane; therefore the motion perpendicular to this direction, *i.e.* parallel to the plane, is unaltered;

$$\therefore v \cos\beta = u \cos a.$$

If the impact is direct, there is no motion parallel to the plane before, and therefore none after, impact.

And also the motion normal to the plane is altered in the same way as if the motion parallel to the plane did not exist. Now the plane being fixed, $u \sin a$, the velocity of the ball in this direction, is the relative velocity of the ball and plane before impact; and $v \sin\beta$ is the relative velocity after impact, if the ball is elastic. Hence in this case $v \sin\beta : u \sin\alpha = e$, Art. 164; $\therefore v \sin\beta = eu \sin\alpha$.

Hence $\tan\beta = e \tan\alpha$, and $v = u \sqrt{\cos^2\alpha + e^2 \sin^2\alpha}$.

But if the ball is inelastic, it does not separate from the plane after impact, and therefore it only moves along the plane, or is at rest, according as the impact was oblique, or direct.

169. The same conclusions are true, if the ball impinges upon any smooth fixed curved surface, only using the phrase " along a tangent to the surface," instead " along the plane," etc.

EXAMPLES.—XXVIII.

(1.) A ball impinges on a plane at an angle 60° (*i.e.* the angle its direction of motion before impact makes with the normal is 30°), moving with a velocity 5. Given that the modulus of elasticity is $\frac{1}{3}$, determine the magnitude and direction of the velocity after impact. What is the angle between its two directions of motion?

(2.) By the impact of a ball on a plane the direction of its motion is turned through a right angle, the modulus of elasticity being ·25; determine the direction of the ball's motion before impact.

(3.) Two points A and B are taken on the diameter of a circle and on opposite sides of its centre O. If OA, OB be respectively one half, and one sixth, of the radius, show that a perfectly elastic ball projected from A at an angle $\cot^{-1}\dfrac{5}{8}$ to the diameter will, after rebounding from the circle, pass through B.

(4.) A billiard ball, moving in a line perpendicular to a cushion, impinges directly on an equal ball at rest at a distance a from the cushion. Show that they will again impinge at a distance $\dfrac{2e^2}{1+e^2}a$ from the cushion, e being the coefficient of elasticity between the balls and the cushion.

(5.) A perfectly elastic particle is projected from a corner of a rectangle in the direction bisecting the angle between the adjacent sides, under what conditions will it return to the point of projection, and how many impacts will have taken place against the sides?

(6.) **An imperfectly** elastic ball is projected along a smooth horizontal table in the direction AO, it strikes a smooth vertical plane at O and rebounds in the direction OB; it is then projected along BO and rebounds in the direction OC. If the angle AOC be the greatest possible, prove that the acute angles of inclination of OA, OB, OC to the verticle plane are in $A.P.$

(7.) Two elastic balls, A and B, of equal radii, the modulus of whose elasticity is $\frac{1}{3}$, and such that $3B=5A$, lie on a smooth horizontal plane. A impinging on B at rest drives it against a vertical wall of same elasticity. Show that B on returning will meet A at a distance from the vertical plane equal to one-third of its original distance.

(8.) A and B are two balls lying on a horizontal table bounded by a straight cushion ; being given the distances of A and B from the cushion and from each other, and the coefficient of elasticity between the ball and the cushion, find the direction in which A must be struck so that after rebounding from the cushion it may hit B.

(9.) ABC is a triangle, and AP, BQ are drawn perpendicular to BC, CA. A perfectly elastic ball is projected from P along PQ, show that after impinging on CA, AB (supposed smooth), it will return to P, and pursue the same path as before.

(10.) ABC is a horizontal circle ; a ball projected from A is reflected at B and C, and returns to A ; show that the time from A to B : the time from C to $A=$ the modulus of elasticity.

170. *Compression and Restitution.*

When one body in motion comes in contact with another, it is found that its surface is bent, and each body compressed about the part where it is in contact.

Thus, if a smooth fixed slab is smeared with a fine coloured matter and a white sphere of ivory is let fall on it, a small coloured spot of finite size is found on the ivory, showing that the ball was in contact, not at a geometrical point, but throughout the small spot; hence the sphere must have been flattened in and the ivory compressed. Further, the ivory ball afterwards presents no appearance of being permanently flattened, and has recovered its original shape. Hence after the compression a restitution of shape took place. The effort towards restitution in such cases brings into play a force which separates the bodies. Hence they are elastic bodies whose shapes are restored.

If, on the other hand, we did the same with a ball of wax, we should find the coloured spot, but it would be flattened, and apparently no restitution would have taken place. Thus between bodies, in which no restitution occurs, after the greatest amount of compression is reached, no separation takes place, *i.e.* they are inelastic. It will be observed that, until the moment of greatest compression, elastic and inelastic bodies behave in exactly the same manner, and that the difference consists in this, that the elastic bodies exert on each other a force of restitution, and the inelastic bodies do not exert such a force.

Now as long as two spheres are being compressed they seem to be, as it were, approaching each other, and it is not until the compression is completed (which, we suppose, takes place at the same instant for each ball) and the restitution begun that they begin to separate. So that at the moment of greatest compression they are neither approaching nor separating, *i.e.* they have a common velocity. Denote it by U. Call the bodies A and A'. Denote their masses by m and m', and their previous velocities in the line of centres by u and u'.

Hence $mu-mU$ and $m'u'-m'U$ denote the changes of momenta which have been produced.

Now let R_1 and $-R_1$ denote the amounts of force which have been expended by A on A', and by A' on A, up to this moment. R_1 is called the force of compression.

These two amounts of force have produced the above changes in the momenta.

Hence $mu-mU=-R_1,\ m'u'-m'U=R_1$;

$$\therefore u-U=-\frac{R_1}{m},\quad u'-U=+\frac{R_1}{m'};$$

K

$$\therefore -\frac{R_1}{m} - \frac{R_1}{m'} = u - u', \text{ and } R_1 = \frac{mm'}{m+m'}(u'-u).$$

Hence, Art. 166, (7), $R = (1+e)R_1$, ... (1).

Again, let R_2 and $-R_2$ denote the amounts of force which A and A' expend on each other after the moment of greatest compression. R_2 is called the force of restitution.

Then $R_1 + R_2 = R$; $\therefore$, from (1), $R_2 = eR_1 = \frac{emm'}{m+m'}(u'-u).$

Note.—In this investigation u and u' denote the components of the velocities of the spheres in the line of centres; therefore R_1 and R_2 are independent of the components perpendicular to the line of centres. Also we assume that the greatest compressions of both balls take place at the same instant. The name of *the coefficient of restitution* is sometimes given to e.

171. We will now give examples of a class of Problems called sometimes *Impact of Projectiles.*

Ex. 1. A particle is dropped from a height a above the

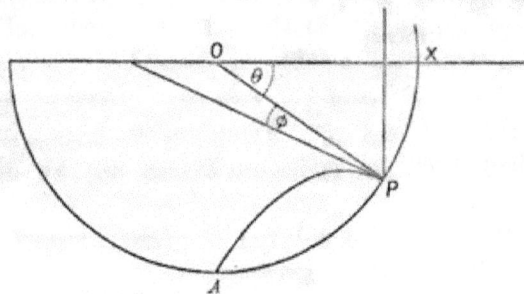

Fig. 44.

centre of a bowl of radius a, the coefficient of elasticity being e. Find where it must hit the bowl in order that after rebounding it may pass through the lowest point A.

Let O be the centre, OX the horizontal radius, P the point of impact. Denote the angle XOP by θ. Then the angle of impact is $\frac{\pi}{2} - \theta.$

Just before impact, the vertical height through which the particle has fallen is $a + a \sin\theta$, and its vertical velocity (u) is given by $u^2 = 2ga(1 + \sin\theta)$ . . . (1), and is equivalent to $u \cos\theta$, perpendicular to OP,

and $u \sin\theta$, along OP.

The former is unaltered by the impact; but the latter is changed to $eu \sin\theta$ along PO, Art. **168**.

Again, *just after* impact, let v be its velocity, ϕ the angle its direction of motion makes with PO.

Hence $v \sin\phi = u \cos\theta$, and $v \cos\phi = eu \sin\theta$; and the angle, which its direction of motion makes with the horizon, is $\theta - \phi$. Therefore its velocity is equivalent to, horizontally, $v \cos(\theta - \phi) = eu \sin\theta \cos\theta + u \sin\theta \cos\theta$

$$= (1 + e)u \sin\theta \cos\theta ; . \quad . \quad (2)$$

and vertically, $v \sin(\theta - \phi) = eu \sin^2\theta - u \cos^2\theta$. . . (3).

Let t be the time of going to P from A. Then in time t the particle traverses $a \cos\theta$ horizontally, and $a - a \sin\theta$ vertically downwards;

$$\therefore a \cos\theta = v \cos(\theta - \phi)t \quad . \quad . \quad . \quad (4),$$

$$-a(1 - \sin\theta) = v \sin(\theta - \phi)t - \tfrac{1}{2}gt^2 \quad . \quad . \quad (5).$$

Now, eliminating t between (4) and (5), and v and ϕ by means of (2) and (3), and u by means of (1), we have the equation

$$4(1 + e)(1 + \sin\theta) \sin\theta \{(1 + e) \sin\theta - 1\} = 1,$$

for determining θ, which defines the point of impact.

Ex. 2. A projectile starts from a point in a smooth horizontal plane, with velocity u, at an angle θ to the horizon,—the co-efficient of elasticity between the projectile and plane being e.

On reaching the plane again it has

a vertical velocity $u \sin\theta$, downwards,

and a horizontal velocity $u \cos\theta$.

After the impact the latter remains the same, and the former is changed to $eu \sin\theta$, upwards.

Just after the next impact the vertical velocity is $e^2u \sin\theta$, and so on.

Also, the intervals, which elapse between the instants when the projectile is on the plane, are $2\dfrac{u\sin\theta}{g}$, $2\dfrac{eu\sin\theta}{g}$, $2\dfrac{e^2u\sin\theta}{g}$, etc.; and the horizontal distances traversed in them are $\dfrac{u\sin2\theta}{g}$, $\dfrac{eu\sin2\theta}{g}$, $\dfrac{e^2u\sin2\theta}{g}$, etc.

Hence the whole horizontal range up to the nth impact is

$$\frac{u\sin2\theta}{g}(1+e+e^2+\ldots+e^{n-1})=\frac{u\sin2\theta}{g}\cdot\frac{1-e^n}{1-e}.$$

The latus rectum of the path during each interval is the same, viz., $\dfrac{2u^2\cos^2\theta}{g}$; and the greatest heights in the intervals are $\dfrac{u^2\sin^2\theta}{2g}$, $\dfrac{e^2u^2\sin^2\theta}{2g}$, $\dfrac{e^4u^2\sin^2\theta}{2g}$, etc.

Ex. 3. A particle is projected and strikes against a vertical wall (coefficient of elasticity=e) at a distance a from the point of starting. Find when and where it will meet the horizontal plane through the point of starting.

Let v be the velocity, and θ the angle, of projection.

Then initially the velocity is equivalent to

$$v\cos\theta, \text{ horizontally,}$$
$$\text{and } v\sin\theta, \text{ vertically.}$$

Let t be the time of reaching the wall;

$$\therefore a=v\cos\theta.t.$$

Now the vertical velocity is unaltered by the impact, therefore the vertical motion is the same, and the particle will reach the horizontal plane in the same time, as if the wall were not there.

Hence if $t+t'$ is the whole time of flight,

$$v\sin\theta-g\frac{t+t'}{2}=0, \text{ or } t+t'=\frac{2v\sin\theta}{g};$$
$$\therefore t'=\frac{2v\sin\theta}{g}-t=\frac{2v}{g}\sin\theta-\frac{a}{v\cos\theta}.$$

Also, the horizontal **velocity is** changed to $eu \cos\theta$ **by the** impact.

Hence the particle will reach the ground at a distance from the wall $= ev \cos\theta t' = \dfrac{2ev^2}{g} \sin\theta \cos\theta - ae = \dfrac{ev^2}{g} \sin2\theta - ae.$

EXAMPLES.—XXIX.

(1.) A mark on a vertical wall appears elevated β degrees above the horizontal plane at a point in it, whence a ball of elasticity e projected at an angle a to the horizon, after striking the mark, rebounds to such point of projection. Prove that $\tan a = (1 + e) \tan\beta$.

(2.) A smooth hemispherical bowl is fixed with its axis vertical, and an elastic particle is dropped into it so as to impinge at a certain point. Find the height from which it must be dropped in order that after rebounding it may just clear the bowl.

(3.) A body is dropped from a height of 100 feet, and rebounds to a height of 60; find the height of its second rebound and the coefficient of elasticity.

(4.) A particle projected with a velocity V, and in a direction inclined to the vertical at an angle a, impinges against a vertical wall whose plane is perpendicular to the plane of its path, and distant h from the point of projection; prove that, if after impact the particle returns to the point whence it started, the elasticity is $\dfrac{gh}{V^2 \sin2a - gh}$.

(5.) A perfectly elastic body is projected upwards from the bottom of an inclined plane, the direction of projection making an angle a with the horizon; and after striking the inclined plane it is reflected vertically upwards. Show that $\tan a = 2 \tan i + \cot 2i$, i being the inclination of the plane to the horizon.

(6.) An imperfectly elastic ball is projected in a direction making an angle of 60° with the horizon, and when at its greatest height is reflected by a vertical plane; determine where the ball will again strike the horizon, and the whole time of flight.

(7.) A plane AB is inclined at an angle β to the horizon, A being its lowest point. A ball (whose elasticity is e) is projected from A in a direction AC making an angle $\left(\dfrac{\pi}{4}-\dfrac{\beta}{2}\right)$ with the plane, and perpendicular to the line in which the plane meets a horizontal plane through A. Find the value of e in order that the ball, after striking the plane, may begin to rise in a direction parallel to AC, and if D, E, F, etc., are the points at which the ball successively rebounds, compare the lengths AD, DE, EF, etc., and find the extreme point in the plane which the ball will reach. Verify the result by finding the perpendicular height to which the velocity of projection is due.

(8.) A perfectly elastic ball is projected from the foot of one of the walls of a room against the opposite wall, in a vertical plane perpendicular to both the walls; show that, if it be required to hit the ceiling after the rebound, the ball must strike the wall at a point at least $\frac{2}{3}$ths of the height of the room from the floor.

(9.) A ball falls from a given height above an elastic smooth plane. Prove that the time of hopping is the same for all inclinations of the plane.

(10.) A ball, of elasticity e, is projected from a point, on a plane inclined at an angle a to the horizon, so as to impinge upon the plane; prove that, if θ, θ_n be the angles which the directions of motion, initially and immediately after the nth rebound, make with the plane,

$$\cot\theta - e^n \cot\theta_n = 2\tan a \frac{1-e^n}{1-e}.$$

MISCELLANEOUS EXAMPLES ON IMPACT.—XXX.

(1.) A ball (of elasticity e) is projected from a point A in a horizontal table AB, at an angle of $45°$, so as to strike a vertical wall through B. The plane in which the ball moves is perpendicular to the wall, and the height of the point where it strikes the wall above AB is $\frac{1}{4} AB$. If, after rebounding from the wall and *once* from the horizontal table, it just reaches the point A, find the value of e.

(2.) A body is projected from a point in a horizontal plane, so as after one rebound from the plane to strike directly against a vertical wall, and after two rebounds to return to the point of projection. Show that, if $\sin\theta$ be the elasticity, $\sin 3\theta = 5\sin\theta - 4$.

(3.) A ball of elasticity e is projected with a velocity v at an angle β with the plane, whose inclination is a, in a plane perpendicular to the intersection of the inclined plane with the horizontal; show that the ball will cease to hop at a distance from the point of projection equal to

$$\frac{2v^2 \sin\beta \sec a}{g} \frac{}{1-e} \left(\cos\beta - \frac{\sin\beta \tan a}{1-e} \right).$$

(4.) A heavy particle is projected with velocity v from a point in an inclined plane whose angle is $\tan^{-1}\frac{1}{2}$, and hits the plane just at the highest point of its own path; show that, if in its rebound it does the like, its elasticity $=\frac{2}{3}$.

Show also that the time, in which it will after successive rebounds attain the limit of its ascent up the plane, is equal to the time in which it will slide down again to its original point of projection, and that this time $=\dfrac{v}{g} \dfrac{3}{\sqrt{2}}$.

(5.) Two particles of the same elasticity are projected at the same instant from points on an inclined plane, with the same velocity, and in directions making the same angle with the plane, but one up and and the other down. Show that the line joining them is always parallel to the plane.

(6.) A perfectly elastic ball is projected vertically upwards with a velocity of twenty feet. A similar ball is simultaneously let fall to meet it. They meet after five seconds. Find their original distance apart, and describe their subsequent motion.

(7.) A perfectly elastic particle is projected horizontally from the top of a tower 100 feet high in a direction perpendicular to the opposite side of the tower, which is 100 feet distant, and after one impact it strikes the ground just under the point of projection; find the velocity of projection.

(8.) A perfectly elastic particle is projected from a point on the perimeter of a perfectly hard fixed ellipse, so that, after striking the curve n times, it returns to the point of projection. Prove that the length of its path is independent of the position of the point from which it starts.

(9.) If, of two equally and perfectly elastic balls, one is projected so as to describe a parabola, and the other is dropped from the directrix so as just to fall upon the first when at its greatest height; determine the position of the vertex of the new parabola, and the effect of the impact on the times at which the balls will reach the ground.

(10.) One perfectly elastic ball begins to slide down an inclined plane at the same instant that another is projected up the plane with the velocity due to the height of the plane; after impact the first ball ascends to the point from which it was let fall, and the second impinges on a perfectly **elastic obstacle at the** foot of the plane. Show that the two balls **must be of** the same weight, and that **they will meet in a** point three-fourths of the way up the plane.

(11.) **A** perfectly elastic ball is projected from **such a point in a** wall of a triangular room **and** in such a direction **that,** after impact at **the two** other walls, it returns to the point of projection, prove that

the least velocity of projection is $2S\sqrt{\dfrac{2g}{abc}}$, a, b, c being the horizontal lengths of the **walls, and** S the area of the floor. **Also with** this velocity find **the least** height of the room that **the problem may** be possible.

(12.) A ball is projected **horizontally from** the top of a **staircase,** each step **of** which is a feet high and c feet broad, with a velocity $=\sqrt{2gnc}$; find from which step it will **first rebound.**

(13.) **Three** perfectly elastic equal balls are resting on a billiard table, **the** sides of which are 20 feet and 12 feet in length, in the straight **line bisecting the shorter sides ; each of the** two extreme balls is 5 feet distant **from** the side nearest it ; determine the angle at which the middle ball must strike one of the others that, after two reflections, it may strike the other ball.

(14.) Two equal balls, one perfectly elastic, the other inelastic, are **dismissed** by the same blow from the top of a flight of uniform steps, so that each falls just on the margin of the first step ; show that the number of steps cleared by the elastic ball in its **successive** flights is the series of successive odd numbers, and that **the** two balls reach the **bottom** of the steps simultaneously.

(15.) An elastic particle is projected from a point in a vertical plane **against** a parallel plane, and after $n+1$ impacts at the latter plane, **and** n at the former, returns to the point of projection. The angle of projection being given, find the velocity of projection.

(16.) A **ball** is projected from a point in a smooth plane inclined at an angle a to the horizon, with the velocity v, in a vertical plane which cuts the inclined plane in a horizontal line, and at an angle β to the horizon. The coefficient of elasticity being e, show that the distance taken by the **ball in** its nth bound in the direction of the line of greatest slope on the inclined plane is

$$\frac{2e^{n-1}(1+e)(1-e^{n-1})}{1-e}\cdot\frac{V^2\sin^2\beta\,\sin a}{g}.$$

(17.) A body is thrown vertically downward from a height h, and rebounds from a horizontal plane. If it just reach the point of projection again, find the velocity with which it impinges upon the plane, and show that its (velocity of projection)$^2 = \dfrac{1 - e^2}{e^2}$ (velocity due to height $h)^2$, e being the modulus of elasticity.

(18.) An imperfectly elastic ball (modulus of elasticity $= e$) impinges directly against another ball equal to it, which is at rest. The second ball strikes a cushion of elasticity e' at a distance a and rebounds; find the place and time at which they will meet again.

(19.) Two imperfectly elastic balls, equal in size, but unequal in mass, are placed between two perfectly hard vertical planes, to which the line joining the centres of the balls is perpendicular, each ball being initially at a distance from the plane nearest to it inversely proportional to its mass. The balls approach one another with velocities inversely proportional to their masses; prove that every impact will take place at the same point as the first does.

(20.) A and B are given positions on a smooth horizontal table; and AC, BD are perpendiculars on a hard plane at right angles to the table. If a ball struck from A rebounds to B after impact at the middle point of CD, show that, when it is sent back from B to A, the point of impact on CD will divide it in parts whose ratio is $e^2 : 1$, when e is the elasticity of the ball.

(21.) A number of balls, whose elasticity is $\frac{1}{2} (\sqrt{2} - 1)$, are let fall on an inclined plane, and each strikes it the second time twice as far down it as the first time; show that the points from which they fall lie in a line perpendicular to the plane.

(22.) Two bodies, P and Q, of which Q is inelastic, and P is perfectly elastic and heavier than Q, are connected by an inextensible string which passes over a smooth fixed pulley. They start from rest at the same distance a from a fixed horizontal plane, and when P impinges on the plane and rebounds with unchanged velocity, Q strikes against a fixed obstacle and is reduced to instantaneous rest; determine the subsequent motion, and show that the two bodies are again at instantaneous rest when P is at a height $\dfrac{P^2 a}{(P + Q)^2}$ above the horizontal plane.

(23.) A, B, C are three equal smooth balls, situated on a horizontal table, and forming an isosceles triangle having an obtuse angle at B; if A be struck so as that having hit B it shall hit C, show that B will move in a direction inclined to AC at an angle ϕ, given by the equation

$$\sin 2\phi = \frac{3-e}{1+e}\sin 2A,$$

e being the elasticity of the balls.

(24.) Two equal scale pans, each of mass M, are connected by a string which passes over a smooth peg, and are at rest. A particle of mass m is dropped on one of them from a height $\frac{u^2}{2g}$, the coefficient of elasticity between the particle and the scale pan being e. Find the velocity of the scale pans after the first impact, and show that if the length of the string exceed $\frac{2eu(1+e)}{g} \cdot \frac{mu}{m+2M}$ a second impact will take place.

Also, prove that if the string be long enough the velocity of the scale pans after the nth impact will be $(1+e)\frac{1-e^n}{1-e}\frac{mu}{m+2M}$, and that the particle will come to relative rest after a time $\frac{2eu}{g(1-e)}$.

(25.) A string, passing over a pulley at the top of an inclined plane, connects two equal particles, one of which is placed on the plane and the other hangs freely; below the descending point is a perfectly elastic horizontal plane; prove that, if the string become stretched when this particle has reached its greatest height after the nth rebound, the inclination of the plane is $\sin^{-1}\dfrac{4n-1}{(2n-1)^2}$.

(26.) Two weights Q, P $(Q > P)$ connected by an inelastic string passing over a smooth pulley, are initially at rest, Q being at a distance s above a horizontal table, the modulus of elasticity between which and Q is e; the system is then allowed to move freely, and the table is removed after Q has impinged upon it. Show that the velocity of P just after the string again becomes tight is equal to $\dfrac{[2gs(Q-P)]^{\frac{1}{2}}(Q-eP)}{(Q+P)^{\frac{3}{2}}}$.

(27.) A parabola is placed in a vertical plane with its axis vertical and vertex downwards. A particle, whose elasticity is $\frac{1}{2}$, strikes the curve at an angle $\sin^{-1}\dfrac{1}{\sqrt 5}$ with velocity due to half the latus rectum. Find where it strikes if after rebounding it passes through the vertex.

(28.) Three equal perfectly elastic individuals start simultaneously, skating on the face of a smooth triangular sheet of floating ice, from the angles of the triangle to meet at the centre of gravity; determine their subsequent motion.

(29.) When a projectile arrives at an end of the latus rectum of its path, another equal body falls upon it from the directrix. Find the consequent change in its path, supposing the bodies perfectly elastic.

(30.) A perfectly elastic particle is dropped from a point on the interior of a smooth sphere; show that, after its second impact on the sphere, it will ascend vertically, and will continually pass and repass along the same vertical and parabolic paths, if the horizontal distance of its first vertical path from the centre be $\dfrac{(3 - \sqrt{2})^{\frac{1}{2}}}{2}$. (the radius of the sphere).

(31.) An imperfectly elastic ball is dropped into a hemispherical bowl from a height n times the radius of the bowl above the point of impact, so as to strike the bowl at a point distant $30°$ of arc from its lowest point, and just rebounds over the edge of the bowl. Find the elasticity of the ball.

(32.) A ball of elasticity e is projected from a given point with a given velocity at an elevation a, and impinges at the highest point of its trajectory against a plane inclined to the horizon at an angle β. If the ball after impact descend vertically, show that $\cot\beta = \sqrt{e}$, and the velocity on reaching the horizontal plane through the point of projection is to the velocity of projection as $\sqrt{e} \cos a + \sin a$ to 1.

(33.) In a game of croquet a ball which is to be croqued is at a certain distance on one side of a hoop; the striker wishes to place his ball so that after the croquet it may be in front of the hoop, and the other ball be at the same distance behind it. Show that the player must give his stroke in the direction of the hoop, and that the line joining the centres of the two balls must be inclined at an angle $\tan^{-1}\sqrt{e}$ to this direction, e being the coefficient of elasticity between the balls.

(34.) The sides of a triangle ABC subtend equal angles at a point O within it. Prove that if from O, three perfectly elastic balls be projected simultaneously with equal velocities in directions AO, BO, CO, produced respectively, they will, after rebounding from the sides, all meet together simultaneously.

(35.) A body of elasticity e slides down a plane AC inclined to the horizon at an angle a. After impinging on a horizontal plane at C, it strikes horizontally a vertical plane at a distance k from C at a height h above C. Show that $ek \tan a = 2h$, and find AC.

172. PROP. *If m_1, m_2, etc. are the masses of a number of particles moving at any instant in the same, or opposite, directions, with velocities u_1, u_2, etc., then the velocity ($\bar{u}$) of the Centre of Gravity of the system at that instant is*

$$\frac{m_1 u_1 + m_2 u_2 + etc.}{m_1 + m_2 + etc.} = \frac{\Sigma(mu)}{\Sigma(m)}.$$

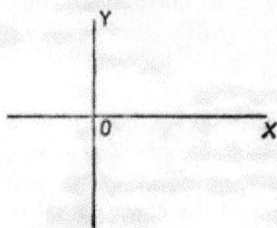

FIG. 45.

Let OY be a line perpendicular to the direction of motion of the particles.

Let x_1, x_2, etc. be the distances of m_1, m_2, etc. from OY at the instant under consideration.

Suppose that the particles retain their velocities for a time t, then at the end of the time t the distances from OY are $x_1 + u_1 t$, $x_2 + u_2 t$, etc.

Let $\bar{x}$, $\bar{x}'$ be the distances of the C.G. from OY at the beginning and end of the time t.

Then $\bar{x} = \dfrac{m_1 x_1 + m_2 x_2 + etc.}{m_1 + m_2 + etc.}$.   .   .   .   .   (1),

$$\bar{x}' = \frac{m_1(x_1 + u_1 t) + m_2(x_2 + u_2 t) + etc.}{m_1 + m_2 + etc.}$$

$$= \frac{m_1 x_1 + m_2 x_2 + etc.}{m_1 + m_2 + etc.} + \frac{m_1 u_1 + m_2 u_2 + etc.}{m_1 + m_2 + etc.} \cdot t$$

$$= \bar{x} + \frac{m_1 u_1 + m_2 u_2 + \text{etc.}}{m_1 + m_2 + \text{etc.}} t \; ;$$

$$\therefore \; \bar{x}' - \bar{x} = \frac{m_1 u_1 + m_2 u_2 + \text{etc.}}{m_1 + m_2 + \text{etc.}} t.$$

Hence as long as the velocities of the particles remain uniform, the change in the position of the C.G. is proportional to the time, and therefore the C.G. moves with uniform velocity, whose measure is $\dfrac{m_1 u_1 + m_2 u_2 + \text{etc.}}{m_1 + m_2 + \text{etc.}}$,

$$\text{or,} \quad \bar{u} = \frac{\Sigma(mu)}{\Sigma(m)}.$$

Cor. 1. If the particles are moving with velocities u_1, u_2, etc. in directions, not parallel to the same fixed line OX, but making with it angles θ_1, θ_2, etc., then the velocities parallel to OX are $u_1 \cos\theta_1$, $u_2 \cos\theta_2$, etc., and we can show, as in the Prop., that the velocity of the C.G.,

parallel to OX is $\dfrac{\Sigma(mu \cos\theta)}{\Sigma(m)}$,

and, perpendicular to OX, $\dfrac{\Sigma(mu \sin\theta)}{\Sigma(m)}$.

Cor. 2. In the expression for the velocity of the C.G. in any direction OX the numerator is the measure of the momentum of the system parallel to OX, and the denominator is the measure of the mass of the system.

Now, in Art. 128, we saw that no action, which takes place amongst the particles of the system, can effect the momentum in any direction, and therefore cannot effect the motion of the C.G. in this direction. And, this being true for all directions, *the motion of the C.G. of a system is wholly unaffected by any action between the parts.*

For example, the motion of the C.G. of two balls impinging is not affected by the impact of the balls.

173. When the velocities of the particles are changing with the accelerations a_1, a_2, etc., of which the directions make angles ϕ_1, ϕ_2, etc. with some fixed straight line OX, then their accelerations parallel to OX are $a_1 \cos\phi_1$, $a_2 \cos\phi_2$, etc. And the velocity of the C.G. parallel to OX at the instant is

$$\frac{m_1 u_1 \cos\theta_1 + m_2 u_2 \cos\theta_2 + \text{etc.}}{m_1 + m_2 + \text{etc.}} = \bar{u}_x,$$

and at the end of an interval t, during which the accelerations remain constant, it is

$$\frac{m_1(u_1 \cos\theta_1 + a_1 \cos\phi_1 t) + \text{etc.}}{m_1 + m_2 + \text{etc.}} = \bar{u}_x + \frac{m_1 a_1 \cos\phi_1 + \text{etc.}}{m_1 + m_2 + \text{etc.}} t.$$

Hence the change in the velocity is proportional to the time; and therefore the *acceleration of the motion of the C.G. parallel to OX is uniform and equal to* $\dfrac{\Sigma(ma \cos\phi)}{\Sigma(m)}$.

Similarly *the acceleration perpendicular to OX is* $\dfrac{\Sigma(ma \sin\phi)}{\Sigma(m)}$.

174. The student must notice that the term *Centre of Inertia* is sometimes used for the point whose position is defined by equations such as (1) of Art. 172. Hence the Propositions in the two preceding Articles might be stated thus: *To determine the velocity and acceleration of the Centre of Inertia of a system, having given the velocities and accelerations of the various parts.*

EXAMPLES.—XXXI.

(1.) Three equal particles are projected, each from an angular point of a triangle, along the sides taken in order, with velocities proportional to the sides along which they move ; prove that their centre of gravity remains at rest. Hence show that, if P, Q, R be points in the sides BC, CA, AB, respectively, of the triangle ABC, such that $\dfrac{BP}{CP} = \dfrac{CQ}{AQ} = \dfrac{AR}{BR}$, then the centre of gravity of the triangle PQR coincides with that of ABC.

(2.) Two equal molecules are connected together by a fine inelastic thread, one of them is placed on a smooth table, the other just over the edge, the thread being at full stretch perpendicular to the edge. Find the velocity of the centre of gravity of the molecules the instant after the former has left the table, and prove that the whole interval of time, from the commencement of the motion to the instant when the thread first becomes horizontal, varies as the square root of the length of the string.

(3.) In the system of pulleys in which each hangs by a separate string, P just supports W; show that, if P is removed and a weight Q substituted, the centre of gravity of the system will descend with acceleration $g.\dfrac{W.(Q-P)^2}{(P^2+QW)(Q+W)}$, the weights of the pulleys being neglected.

(4.) Two equal and perfectly elastic balls, of which the centres are initially at the opposite ends of a diagonal of a rectangle, proceed towards one of the other angles, with velocities proportional to the sides along which they move; determine the direction of motion of each ball after impact, and the path of their centre of gravity.

(5.) Two weights, P and Q, are connected by an inextensible string, the length of which $= l$; and the mass of Q is double that of P. Q is laid upon a perfectly smooth horizontal table (the height of which from the ground $= 2l$), at a distance l from the edge, and P just hangs over the edge. Find the whole time which elapses before P strikes the ground, and compare the velocity of P at that moment with its velocity when Q leaves the table.

Trace the whole path described by the centre of gravity of P and Q.

(6.) In a wheel and axle, the arm of the power P is r, and that of the weight W is r_1. Show that the centre of gravity of P and W moves with the constant acceleration $\dfrac{(Pr-Wr_1)^2}{(P+W)(Pr^2+Wr_1^2)}$.

175. WE have seen, Art. 69, that, if a particle moves in a curve, then, at any point where its velocity is v and the radius of curvature ρ, the normal acceleration is $\dfrac{v^2}{\rho}$. If m is the mass of the particle, the force necessary to produce this acceleration is $\dfrac{mv^2}{\rho}$. So that the resolved part, in the direction of the normal, of the resultant force acting on the particle, *i.e.* the sum of the resolved parts in the direction of the normal of all the forces acting, is $\dfrac{mv^2}{\rho}$.

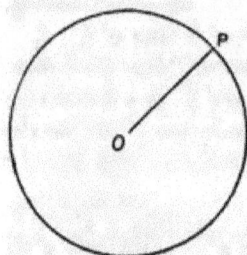

176. *Ex.* 1. Thus if a particle P, of mass m, be tied by a string of length a to a fixed point O in a smooth horizontal plane, on which the particle moves uniformly with velocity v in a circle, the string being tight, then all the string will have to do is to produce this necessary normal force, and we shall have the tension $= \dfrac{mv^2}{a}$.

FIG. 46.

Ex. 2. Again, let a be the equatorial radius of the earth, ω its angular velocity, P the pressure which the earth exerts on a body of mass m placed at the equator, mf the force of the earth's attraction, then $mf - P = m.a\omega^2$;

$$\therefore P = mf - ma\omega^2.$$

So that the pressure between the earth and the body, or the apparent weight of the body, is less than if the earth were at rest by the amount $ma\omega^2$.

Also $P = mg$; $\therefore g = f - a\omega^2$.

160

Note. Here ω is the angle described by any equatorial radius in one second $=\dfrac{2\pi}{24 \times 60 \times 60}$.

Ex. 3. Let a particle P, of mass m, be hung from a fixed point O by a string of length l, and be made to revolve in a horizontal circle of radius a.

Let OA be the vertical line through O, PA perpendicular to OA. Then A is the centre of the circle and $AP = a$.

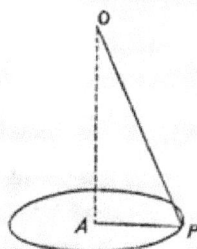

Let $AOP = \theta$, so that $\sin\theta = \dfrac{a}{l}$.

Fɪɢ. 47.

Let T be the tension of the string, and v the velocity of P. Then T is equivalent to $T \cos\theta$ vertically,

and $T \sin\theta$ along PA.

Now there is no vertical motion;

$$\therefore T\cos\theta - mg = 0, \text{ or } T = mg \sec\theta.$$

Also the necessary normal force, along PA, is produced by the component $T \sin\theta$;

$$\therefore T\sin\theta = \frac{mv^2}{a}.$$

Hence $\tan\theta = \dfrac{v^2}{ag}$, or $v^2 = ag \tan\theta$.

177. We can now explain why on a railway curve the permanent way is tilted so that the outside wheels of a carriage travelling on it are highest.

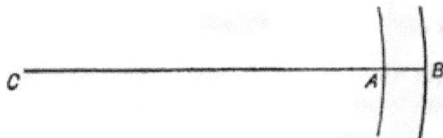

Fɪɢ 48.

Let m be the mass and v be the velocity of the carriage, and ρ the radius of curvature at any point.

L

Then, in order to make the carriage keep on the rails, there has to be a force in the plane of the curve along the normal inwards $= \dfrac{mv^2}{\rho}$.

Now, if the cross section of the way is horizontal at that point, this force has to be produced by the pressure of the rails sideways on the wheels, which therefore must be equal to $\dfrac{mv^2}{\rho}$.

[As a matter of fact, this pressure is produced by the outside rail pressing by means of its inner side against the flanges of the wheels running on it.]

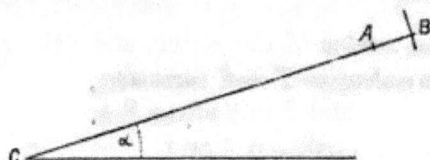

FIG. 49.

But, if the way is tilted to an angle a with the horizon, then the weight mg has a component $mg \sin a$ along the normal inwards, so that now the force $\dfrac{mv^2}{\rho}$ is made up by the pressure and the $mg \sin a$; ∴ the pressure need only be $\dfrac{mv^2}{\rho} - mg \sin a$.

Thus the more the way is tilted the more the sideways pressure of the rails is relieved; or, if we like, the carriage may go faster, and yet only require the same pressure. Thus, let v' be the new, and greater, velocity, then $\dfrac{mv'^2}{\rho} - mg \sin a = \dfrac{mv^2}{\rho}$,

or $v'^2 = v^2 + g\rho \sin a$.

Hence, on account of the tilt, the carriage is more easily kept on the rails. If at any point the sideway pressure should, from the weakness of the rails or any other cause, be less than the required amount, the carriage would leave the rails towards the outside of the curve.

178. *Centrifugal Force.*—It would seem that formerly it used to be expected that a body would move in any path arbitrarily assigned to it.

Now, if a particle is moving in a curve, when at any point it is moving along the tangent at the point, and, when at a consecutive point it is moving along the tangent at that point; hence its direction has been changed during its passage from the first to the second point, and we know, Art. **175**, that the force which must be applied to make this change is $\dfrac{mv^2}{\rho}$ along the normal inwards. If this force were not applied the particle would continue to move in the tangent at the first point, that is, it would quit the curve.

When it was found that this force had to be applied simply for the purpose of retaining the particle in its arbitrarily assigned path, it seems as if it was supposed that the particle exerted on itself a force, on account of which it had a tendency to fly out of the curve (*i.e.* away from the centre, in the case of the most ordinary kind of curvilinear motion, viz., circular), and which had to be counteracted by the inwards normal force $\dfrac{mv^2}{\rho}$ before the particle would move on the curve.

Centrifugal force was the name given to this outwards normal force, which was supposed to have been thus discovered to be exerted by the particle on itself, and it was supposed to be exerted in the direction of the normal outwards and to be equal to $\dfrac{mv^2}{\rho}$. This exertion of a force by a particle on itself would be contrary to the First Law of Motion.

At the present day this language is still often retained. For instance, it is said that, " on a carriage going round a railway curve, centrifugal force would press the outside wheel against the outer rail."

EXAMPLES.—XXXII.

(1.) What alteration of the absolute attraction of the earth's mass, the length of day remaining unchanged, would make the weight of a body zero at the equator?

(2.) The breadth between the rails in a railway is 4 feet 8½ inches. Show that on a curve of 500 yards radius, the outer rail ought to be raised about 2⅛ inches for trains travelling thirty miles an hour.

(3.) A railway train is moving with a uniform velocity v on a curved line in a horizontal plane, and the friction is $\frac{1}{n}$th of the weight. Show that, if the wheels of any carriage (whose mass is m and length a) exert no sideways pressure on the rails, the tension of its fastenings is approximately equal to $\frac{mv^2}{a}$; and prove that this will nearly be the case at a carriage whose number from the end of the train (if there be enough) is the nearest integer to $\frac{nv^2}{ag}$.

(4.) Calculate the time in which the earth should revolve so that a body at the equator should fall 14 feet in a second.

(5.) A string two yards long, passing through a hole in a smooth table, has four pounds weight suspended by it, and one pound weight attached to the other end and resting on the table at a distance equal to two thirds the whole length of string from the hole. With what velocity must the latter weight be made to revolve, in order that the other may be supported at rest?

(6.) Determine the velocity with which the earth (supposed to be a sphere) must rotate in order that bodies within 30° of the equator may begin to leave it.

(7.) Two equal particles are attached at points B and C of a string, which is itself attached to a fixed point A, and the particles are projected so as to describe horizontal circles uniformly; show that, if the portions of the string are equal, and a, β are the angles made by them with the vertical,

$$\frac{2\tan a}{\tan \beta} = \frac{2\sin a + \sin \beta}{\sin a}.$$

(8.) Supposing a body fastened to a string, which can just bear without breaking a weight of 12 lbs., and swung round horizontally in a circle with a radius of 3 yards, 160 times a minute; find the greatest weight of the body consistent with the strength of the string.

179. WHEN a body is acted on by a force and is moving so that the point of application of the force is displaced, wholly or in part, in the line of action of the force, then *Work* is said to be *done* on the body. If the displacement is *in* the direction of the force, the work is said to be done *by* the force. If it is *opposite* to the direction of the force, the work is said to be done *against* the force.

Thus when coals are hauled up a pit, work is done on them, by the tension of the rope, and against the weight.

Again, if a particle is moving on an inclined plane *AB*, and is pulled by a rope along the plane upwards, then, in moving

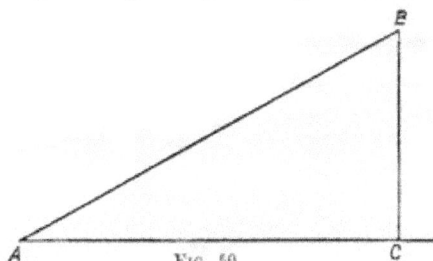

FIG. 50.

from *A* to *B*, work is done on the particle by the tension and against the weight, but no work is done by the pressure of the plane. For, draw *AC* and *BC* horizontally and vertically, respectively. The particle goes, through a distance *AB* in the direction of the tension, through a distance *CB* opposite to the direction of the weight, and through no distance in the direction of the line of action of the pressure for it always moves perpendicular to that line.

165

It should be noted that, if the particle had been pulled up
CB, the *same* amount of work would have been done against
the weight; for the force would have been the *same* in magnitude, and the particle would have gone through the *same*
distance in the line of action of the force, and *opposite* to the
direction of the force.

180. PROP. *The measure of the work done by a force is F.h, where
F is the measure of the force and h the measure of the distance
which the point of application goes in the direction of the force.*

For we take as our unit the work done by a unit of force as
its point of application moves through a unit of distance in its
direction.

Then as the point of application of each of the *F* units, of
which the force is made up, moves through each unit of the
distance, there is a unit of work done by each unit of force,
and therefore *F* units of work by the *F* units of force, and
therefore in the *h* units of distance there will be *F.h* units of
work done.   That is, the measure of the work done is *F.h*.

181. If the work is done *against* the force, the distance,
through which the point moves in the direction opposite to
that of the force, is denoted by a negative number.

For example, a bullet of mass *m* travels along its path from

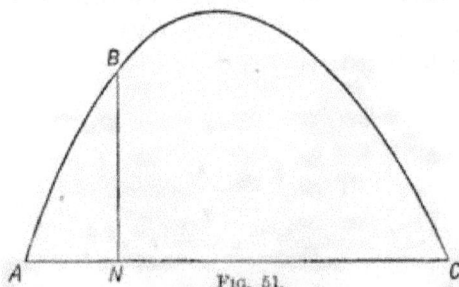

Fig. 51.

A to *B*.   Draw *BN* perpendicular to the horizontal line *AC*
through *A*, and let *h* be its measure.

Then the work done against the weight in moving from A to B is $mg(-h) = -mgh$; and the work done by the weight in moving from B to C is mgh.

182. If a body of mass m is moving with a velocity v, then the product $\frac{m.v^2}{2}$ is called the *Kinetic Energy* (or sometimes the *Vis Viva*) of the body.

If the different parts, of masses m_1, m_2, etc., of a system be moving respectively with velocities v_1, v_2, etc., then $\frac{m_1v_1^2}{2} + \frac{m_2v_2^2}{2} +$etc., is the Kinetic Energy of the system.

183. If a particle of mass m starts with a velocity u, and, whilst travelling through a distance s, is acted on by a force F in the direction of its motion, and, if v is its velocity at the end, then $mv^2 = mu^2 + 2Fs$.

Now the change in the Kinetic Energy $= \frac{mv^2}{2} - \frac{mu^2}{2}$, and the work done by the force is $F.s$. Hence the *Change in Kinetic Energy is equal to the work done.*

Again, in Art. **181**, if s, s' be the distances of A and B from the directrix, the Kinetic Energies at A and B are $\frac{m.2gs}{2}$ and $\frac{m.2gs'}{2}$. Therefore the change in the Kinetic Energy is $mgs' - mgs = -mgh$, which is the work done.

Hence, again, the change in Kinetic Energy is equal to the work done. The student will find this result in all cases of bodies moving under the action of forces.

184. PROP. *To calculate the change in the Kinetic Energy of a system of two balls produced by their direct impact.*

Let m, m' be their masses.

u, u'	,,	velocities before impact.
v, v'	,,	,, after ,,
e	,,	coefficient of elasticity.

Then $v = u - (1+e)\dfrac{m'}{m+m'}(u-u')$

$v' = u' + (1+e)\dfrac{m}{m+m'}(u-u').$

Therefore the Kinetic Energy after impact, *i.e.* $\dfrac{mv^2}{2} + \dfrac{m'v'^2}{2}$,

$$= \frac{mu^2}{2} + \frac{m'u'^2}{2} - (1+e)\frac{mm'}{m+m'}(u-u')^2 + \frac{(1+e)^2}{2}\frac{mm'^2+m'm^2}{(m+m')^2}(u-u')^2$$

$$= \text{Kinetic Energy before impact} - \frac{1-e^2}{2}\cdot\frac{mm'}{m+m'}\cdot(u-u')^2.$$

Now $\dfrac{1-e^2}{2}\cdot\dfrac{mm'}{m+m'}\cdot(u-u')^2$ is never negative, and is always positive except when $e=1$. Hence Kinetic Energy is always lost unless the balls be perfectly elastic.

185. In some books the product mv^2 is called the Vis Viva.

For further information on the subject of this Chapter see *An Elementary Exposition of the Doctrine of Energy*, by D. D. Heath, M.A. London. 1874.

EXAMPLES.—XXXIII.

(1.) Two balls, of weights two pounds and three pounds, are connected by a stretched string passing over a pulley. Determine the amount of work done on the system whilst each ball travels over three feet. What work is done on each ball?

(2.) Determine the change in the Kinetic Energy of the system in Art. 131 when Q begins to move.

(3.) Determine the change in the Kinetic Energy of the system in Art. 132 when R begins to move.

(4.) A force of 15 lbs. pulls a weight of 3 lbs. up an inclined plane of 10 feet length, and elevated at an angle of 30°. Determine the work done, (1) by the force, (2) against the weight of the body.

(5.) During an interval of time, t, an engine, in order to keep up in a train a constant velocity, v, has to do an amount of work, w, against the retarding forces arising from the pressure of the air, friction, etc. Find the average resultant of these forces.

(6.) If a ball moving with a velocity $= V$ impinge directly on a ball of double its mass which is at rest, show that it is brought to rest, and that the other ball moves with a velocity $= \dfrac{V}{2}$; it being premised that the elasticity e is given $= \frac{1}{2}$. What alteration is made in the Vis Viva?

(7.) The force of the steam on the piston of a locomotive does an amount of work, w, during each complete oscillation, and keeps up a uniform motion. Find the average resultant of the retarding forces, the radius of the driving wheel being r.

186. IN this chapter we shall discuss the motion of a heavy particle in some curved path acted on by a force, called the force of constraint, originating in some material body, which constrains the particle to move in this particular path instead of in the straight line, or parabola, in which it would otherwise move.

Thus (1) when a bead slides on a wire bent into the form of a circle, or other curve, it is constrained to move in this circle, or curve, by the pressure of the wire.

Or (2) a particle may slide in a groove, or tube, of small section, the axis of which forms a curve, then the particle would be constrained to move in this curve by the pressure of the substance in which the groove, or tube, is formed.

Again (3) when a body is tied to a string and swung round in a circle, it is constrained to move in this circle by the tension of the string.

In such cases as (1) and (2) we call the curve *a material curve*, and the pressure of the wire or other substance forming the curve is called, for shortness, *the pressure of the curve*, and acts at every point in the direction of the normal to the curve at the point.

We shall only consider cases in which the particle is acted on by gravity and this force of constraint.

We are unable to determine in this treatise a formula connecting any interval of time with the length of arc, which the particle traverses in the interval.

We shall, however, be able to determine the change in the square of the velocity (and therefore the change in the Kinetic Energy) in passing from one point to another at a given vertical distance from the first, and also in some cases the times in which the particle will pass over certain specified arcs.

187. PROP. *An inelastic heavy particle is moving on a smooth curve, whose plane is vertical. Given the velocity at any point, to find its velocity at any other point.*

[Any curve may be considered as formed by a succession of straight lines, the angle between each successive pair being the same, of which ultimately the lengths are endlessly decreased and the number endlessly increased, the above constant acute angle being endlessly decreased. The curve is said to be the limit of this succession of lines, and in all our investigations is supposed to have finite curvature at every point.]

Let ABC be the curve; u the velocity at some point A, v at any other point B; h the vertical height between A and B.

We will suppose A to be above B.

Let the curve AB be the limit of the succession of lines

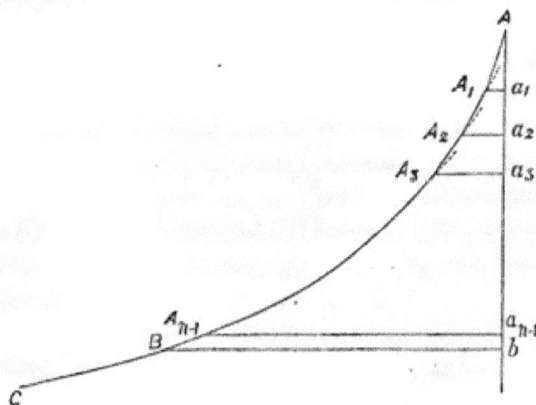

FIG. 52.

$AA_1, A_1A_2, \ldots A_{n-2}A_{n-1}, A_{n-1}B$; the angle between each successive pair being a.

Let the particle start from A, with velocity u, and slide along this succession of lines to B.

Denote the velocities of *arrival* at $A_1, A_2, \ldots A_{n-1}, B$, by $v_1, v_2, \ldots v_{n-1}, v_n$.

Draw Ab vertically through A; and $A_1 a_1, \ldots Bb$ horizontally through $A_1, \ldots B$, to meet Ab in $a_1, a_2, \ldots b$.

Then, Aa_1 being the vertical height of A above A_1,
(Art. 136, 2°) $v_1{}^2 = u^2 + 2g.Aa_1.$.   .   . (1).

Now when the particle arrives at A_1, it impinges, with velocity v_1, against $A_1 A_2$. Resolve v_1 into the two components $v_1 \cos\alpha$ along, and $v_1 \sin\alpha$ perpendicular to, $A_1 A_2$. Of these, $v_1 \sin\alpha$ is wholly destroyed by the impulsive pressure of $A_1 A_2$ brought into play by the impact; and the particle being inelastic, no new velocity in the opposite direction is generated. Also, since this pressure is perpendicular to $A_1 A_2$, no change is produced by the impact in the other component.

Hence the particle starts along $A_1 A_2$ with a velocity $v_1 \cos\alpha$. Now v_2 being its velocity when it arrives at A_2, and $a_1 a_2$ being the vertical height of A_1 above A_2, we have, Art. **136**,

$$v_2{}^2 = v_1{}^2 \cos^2\alpha + 2g.a_1 a_2 \quad . \qquad . \qquad . \qquad . \qquad . \quad (2).$$

Similarly $v_3{}^2 = v_2{}^2 \cos^2\alpha + 2g.a_2 a_3 \quad . \qquad . \qquad . \qquad . \qquad . \quad (3),$

etc. $=$ etc.

$$v_{n-1}^2 = v_{n-2}^2 \cos^2\alpha + 2g.a_{n-2} a_{n-1} \quad . \qquad . \qquad . \quad (n-1),$$

$$v_n{}^2 = v_{n-1}^2 \cos^2\alpha + 2.g.a_{n-1} b \quad . \qquad . \qquad . \qquad . \quad (n).$$

Adding equations (1), (2), . . . (n), and transferring $v_1{}^2 \cos^2\alpha, \ldots v_{n-1}^2 \cos\alpha$ to the left-hand side, we have

$$(v_1{}^2 + v_2{}^2 + \ldots + v_{n-1}^2) \sin^2\alpha + v_n{}^2 = u^2 + 2g.Ab$$
$$= u^2 + 2g.h.$$

Let v' be the greatest of the numbers $v_1, v_2, \ldots v_{n-1}$, then $(v_1{}^2 + v_1{}^2 + \ldots + v_{n-1}^2) \sin^2\alpha$ is not greater than $(n-1)v'^2 \sin^2\alpha$.

Now the angle between AA_1 and $A_1 A_2$ is α,

and   ,, $A_1 A_2$,, $A_2 A_3$ is α;

∴   ,, AA_1 ,, $A_2 A_3$ is 2α.

Also the angle between A_2A_3 and A_3A_4 is α.

$\therefore$ „ AA_1 „ A_3A_4 is 3α; and so on, till finally,   „ AA_1 „ $A_{n-1}B$ is $(n-1)\alpha$.

Denote by ϕ the angle between the tangents to the curve at A and B.

Now when n is endlessly increased and the succession of lines approaches endlessly near to the curve AB, then the lines $AA_1, A_1A_2, \ldots A_{n-1}B$ become the tangents at $A, A_1, \ldots A_{n-1}$; and therefore $(n-1)\alpha$ is then the angle between the tangent at A and A_{n-1}.

Therefore ϕ differs from $(n-1)\alpha$ by the angle between the tangents at A_{n-1} and B; but $A_{n-1}B$ being endlessly diminished, this angle must be so also.—Newton, Lemma vi.

Hence $\phi = (n-1)\alpha$;

$$\therefore (n-1)v'^2 \sin^2\alpha = \frac{\phi}{\alpha}v'^2 \sin^2\alpha = \phi.v'^2\frac{\sin\alpha}{\alpha}.\sin\alpha,$$

and this $=0$ in the limit, since then $\dfrac{\sin\alpha}{\alpha} = 1$ and $\sin\alpha = 0$.

Hence, *à fortiori*, $(v_1^2 + v_2^2 + \ldots + v_{n-1}^2)\sin^2\alpha = 0$.

Also v_n, being the velocity of arrival at $B, = v$.

Hence $v^2 = u^2 + 2gh$.

188. If B had been above A, we should have had $v^2 = u^2 - 2gh$.

189. Suppose that in the above Prop. instead of the phrase "whose plane is vertical," we had had "whose plane is inclined to the horizon at an angle θ."

Then instead of the acceleration g in the vertical direction, we should have had an acceleration $g\sin\theta$ in the direction of the lines of greatest slope. (See Art. 136.)

We should have drawn Ab parallel to the lines of greatest slope, and denoted Ab by h', and made a corresponding change in the phraseology throughout the proof.

Thus we should have arrived at the result

$$v^2 = u^2 \pm 2g\sin\alpha.h'.$$

Now let h be the *vertical* height of A from B, and therefore of A from b, since B and b are in the same *horizontal* line.

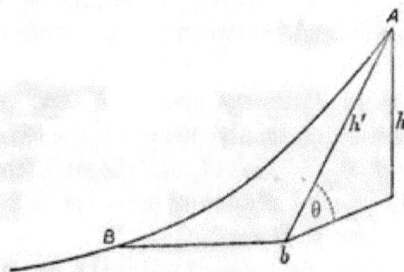

Fig. 53.

Hence $h = h' \sin a$; and $v^2 = u^2 \pm 2.g.h$.

Compare Arts. **146** and **158**.

190. In Art. **187** the particle experiences at every angular point a small impulsive force perpendicular to the line which the particle there begins to slide on. For instance, its amount at A_r was $mv_r \sin a$, m being the mass of the particle.

Now in the limit this line becomes the tangent to the curve at that point, and therefore the direction of the force is normal to the curve at the point.

Also, since the angular points approach indefinitely near to one another, the impulses follow indefinitely closely one after another; and, since a is endlessly decreased, the amount of these impulses is endlessly diminished. Hence we have a succession of indefinitely small impulses following one another at indefinitely small intervals; but such a succession constitutes a continuous finite force.

Hence, when a particle moves on a curve, we have a finite force always acting at **every** point in the direction of the normal.

Again, in passing from one angular point to the next, the particle experiences a pressure perpendicular to the line, and this in the limit becomes a force acting at every point along the normal.

These two forces together make up the pressure of the curve which acts normally at every point.

191. The force of constraint in Art. **189** had better be considered by means of its two components—

(1.) The pressure of the plane, perpendicular to the plane, and equal to $mg\cos\theta$, which compels the particle to move on the plane.

(2.) The pressure of the curve, normal to the curve, which compels the particle to move on the curve instead of on any other line in the plane.

192. In Arts. **187** and **189**, suppose the particle descends from A to B, then $v^2 - u^2 = 2gh$;

$$\therefore \frac{mv^2}{2} - \frac{mu^2}{2} = mg.h.$$

The point of application of the weight, mg, has been transferred from A to B, *i.e.* through a distance h in the direction of the force's action; therefore the work done by it is $mg.h.$

Also there has been no work done by, or against, the pressure of the plane, a force constant in magnitude and direction, since in passing from A to B there is no motion in the direction of this force.

And the same is true of the pressure of the curve; for though the force is of varying magnitude and direction, yet the particle in moving from any position to the consecutive one has no motion in the direction which the force at that instant has. Hence no work is done there, and therefore on the whole no work is done by, or against, this force.

Hence the whole work done in the passage from A to B is $mg.h.$

And $\frac{mv^2}{2} - \frac{mu^2}{2}$ is the change in the Kinetic Energy in passing from A to B.

Hence again we see that the work done in any passage is equal to the change in Kinetic Energy.

193. *Motion in a Vertical Circle.*

We will suppose that the particle is a smooth bead, of mass m, sliding on a wire, or in a tube, forming a circle.

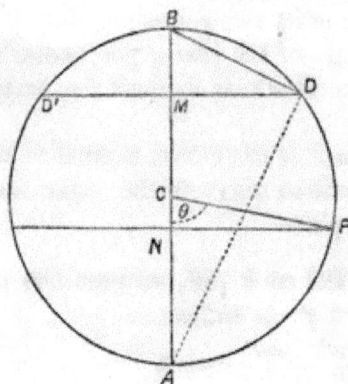

Fig. 54.

Let AB be the vertical diameter, C the centre, and a the radius of the circle.

Let u be the velocity at the lowest point A, v at any other point P.

Let θ denote the angle ACP, and draw PN perpendicular to AB. So that $AN = a - a\cos\theta$.

Then, Art. **188**, $v^2 = u^2 - 2g.AN$ . . . . (1)

$$= u^2 - 2ag + 2ag\cos\theta.$$

Let R be the pressure of the curve at P, inwards, along PC.

Then the sum of the resolved parts of the forces acting on the particle at P in the direction of the normal is $R - mg\cos\theta$;

$$\therefore, \text{Art. 175, } R - mg\cos\theta = m\frac{v^2}{a} = m\left(\frac{u^2}{a} - 2g + 2g\cos\theta\right);$$

$$\therefore R = m\left(\frac{u^2}{a} - 2g + 3g\cos\theta\right).$$

Hence when P is at A, $\theta = 0$, and $R = m\left(\frac{u^2}{a} + g\right)$, and, as P moves from A towards B, θ increases, and therefore R diminishes.

Also, $R=0$, if $\dfrac{u^2}{a}-2g+3g\cos\theta=0$, *i.e.* if ever the particle

comes to the point where $\cos\theta=\dfrac{2ag-u^2}{3ag}$; and if the particle

goes beyond this point, R becomes negative, and the pressure of the curve is outwards, along CP.

We shall have different kinds of motion according as $u^2 < = $ or $> 2g.AB$.

1°. If $u^2 < 2g.AB$, let M be the point in AB such that

$$u^2 = 2g.AM.$$

Draw DMD' horizontally to meet the circle in D and D'. Then when the particle arrives at D it stops, since by (1) $v=0$ there, and then begins to descend. Since it starts from D with no velocity, when it arrives at A the square of its velocity $=2g.AM=u^2$.

Hence it will rise on the other side of A to the same height AM, and therefore at D' will again be at rest, and will thus continue to oscillate between D and D'.

Now at D, since $v=0$, $u^2-2ag+2ag\cos\theta=0$;

$$\therefore \cos\theta=\dfrac{2ag-u^2}{2ag}.$$

This gives the angle ACD.

[If $u^2=2ag$, the particle just rises to the end of the horizontal diameter, and oscillates through the lower half of the circle.]

Join AD. Then, from the right-angled triangles BDA, AMD, $AD^2 = AM.AB$.

$$\therefore u^2 = 2g.AM = \dfrac{2g}{AB}\cdot AD^2, \text{ or } u = \sqrt{\dfrac{2g}{AB}\cdot AD} = \sqrt{\dfrac{g}{a}}\cdot AD.$$

Hence, when a particle oscillates in a circle, the velocity at the lowest point varies as the chord of half the arc through which it oscillates.

2°. If $u^2=2g.AB$, the particle just rises to B and stops there.

M

3°. If $u^2 > 2g.AB$, $u^2 - 2g.AN$ cannot vanish, for the greatest possible value of AN is AB.

Hence the particle never stops, but arrives at B with a velocity $\sqrt{u^2 - 2g.AB}$, and descends the other side, and arrives at A again with a velocity whose square is, Art. 187,

$$(u^2 - 2.gAB) + 2g.AB = u^2.$$

And thus the motion is kept up perpetually, the particle always going round in the same direction.

194. Suppose now instead of moving in a tube, or on a wire, the particle is tied to one end P of a string, of length a feet, of which the other end C is fixed, and thus moves in a vertical plane. As long as the string remains stretched the particle describes a circle whose centre is C.

The string by its tension (R) is capable of exerting a normal force on the particle in the same manner as the tube, or wire, did; except that it can only do this inwards, along PC.

Hence the particle, if its velocity at its lowest point A is u, will move exactly in the same manner as in Art. 193, until it reaches the point where the pressure became zero, and afterwards acted outwards. If the particle does not stop at this point, the string will not be able to exert the necessary force outwards to retain it in the circle, and it will move within the circle, and the string will become slack. The only force now acting on the particle is gravity; hence it will describe a parabola until the string again becomes tight.

We will now investigate the conditions for the string remaining stretched.

If the particle remain on the circle, it does not stop till it comes to D, where $\cos ACD = \dfrac{2ag - u^2}{2ag}$.

And the tension vanishes at the point E where $\cos ACE = \dfrac{2ag - u^2}{3ag}$, and beyond this point, if the particle were to remain on the circle, the tension would have to be negative.

Now $\cos ACE <$, or $=$, $\cos ACD$, numerically, and therefore ACE is nearer $90°$ than ACD is, unless $ACE = ACD$.

If $u^2 < 2ag$, both $\cos ACD$ and $\cos ACE$ are positive; $\therefore$ both ACD and ACE are < 90; $\therefore ACD < ACE$, and the particle stops before the tension vanishes, and $\therefore$ continues to oscillate through an arc $= 2ACD <$ a semicircle.

If $u^2 = 2ag$, both $\cos ACD$ and $\cos ACE$ vanish; therefore $ACD = ACE = 90°$; and the particle just rises to the end of the horizontal diameter, and the tension just vanishes but does not require to be negative, and the particle oscillates through a semicircle.

If $u^2 > 2ag$ and $=$, or $<$, $4ag$, both $\cos ACD$ and $\cos ACE$ are negative, $\therefore$ both ACD and ACE are > 90; $\therefore ACE < ACD$; and the tension requires to be negative before the particle stops, and therefore the particle leaves the circle.

If $u^2 > 4ag$, $\cos ACD$ is numerically > 1, and therefore there can be no such point as D, i.e. no stopping point.

Now if $u^2 < 5ag$, there is a point E where the tension vanishes, and beyond E the particle leaves the circle.

But if $u^2 > 5ag$, $\cos ACE$ is numerically > 1, and there is no point where the tension vanishes. Therefore the particle makes complete revolutions round the circle.

Hence the string remains stretched, if $u^2 < 2ag$, or $> 5ag$.

The greatest tension which the string has to bear is $\frac{m}{a}(u^2 + ag)$, its value when P is at the lowest point.

Now, if P oscillates in the circle, u^2 must not be greater than $2ag$, therefore the string need not be capable of bearing a greater weight than $\frac{m}{a}(2ag + ag)$, i.e. $3mg$; but if P goes completely round the circle, u^2 must be at least $5ag$, therefore the string must be capable of bearing at least $\frac{m}{a}(5ag + ag)$, i.e. $6mg$.

If we had an open groove cut *inside* a circle, the motion would be exactly the same as in the case of the string.

195. A particle, of mass m, moves in a groove cut on the *outside* of a vertical circle, of radius a, and is projected from the highest point B, with a velocity u. To find where it will quit the circle.

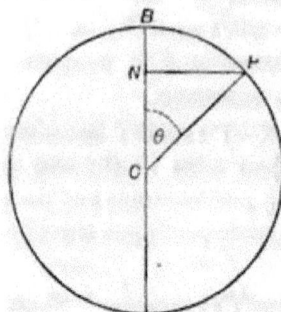

Here the groove is only able to exert an outward normal pressure, hence the particle will quit the circle as soon as it requires **an inward normal force of constraint to keep it on the circle.**

Let C be centre, P some position of the particle before quitting the circle. Denote the angle BCP by θ, and the velocity of P in this position by v.

Fig. 55.

Then, Art. **187**; $v^2 = u^2 + 2(a - a\cos\theta)g$.

Let R denote the pressure on P *outwards*. Then $mg\cos\theta - R$ is the sum of the resolved parts of the forces along the normal inwards; $\therefore$, Art. **175**, $\dfrac{mv^2}{a} = mg\cos\theta - R$;

$$\therefore R = \frac{m}{a}(ag\cos\theta - v^2) = \frac{m}{a}(ag\cos\theta - u^2 - 2ag + 2ag\cos\theta)$$

$$= \frac{m}{a}(3ag\cos\theta - 2ag - u^2).$$

Hence, R will vanish when $\cos\theta = \dfrac{u^2 + 2ag}{3ag}$; and after this

point R would have to be negative to retain the particle on the circle.

Therefore, on leaving this point the particle quits the circle, in the direction of the tangent to the circle at this point, with

a velocity $= \sqrt{u^2 + 2ag - 2ag\dfrac{u^2 + 2ag}{3ag}} = \sqrt{\dfrac{u^2 + 2ag}{3}}$;

and proceeds to describe a parabola.

196. We can now explain the method for determining the coefficients of elasticity of bodies, referred to in Art. **162.**

This method was originally devised by Newton, and the experiments have since his time been repeated by various persons. In 1834 Mr. Hodgkinson presented to the British Association a report of his researches into the subject. Of this report we have availed ourselves in the following explanation.

197. Two balls, A, B, are hung from points, C, D, with *equal* radii (*i.e.* equal distances from C and D to the centres of the balls) so as just to be in contact when the strings are vertical; and the curves AEH, BFG are circular arcs, round the centres C, D, at the lowest points of which the balls are when the strings are vertical.

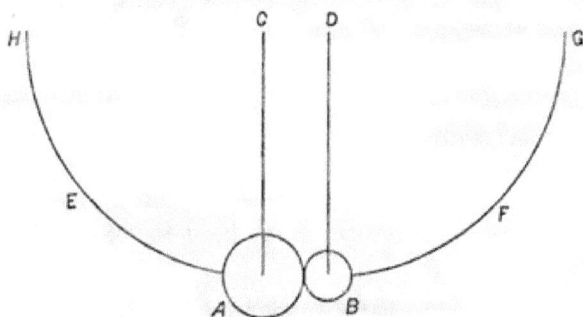

Fig. 56.

Now, Art. 193, 1°, if either ball, say B, is drawn aside and then let fall, its velocity on arriving at the lowest point, *i.e.* when BD is vertical, is proportional to the chord of the arc through which it has fallen; and, if either start from its lowest point, its velocity at starting is proportional to the chord of the arc through which it ascends before coming to rest. This result was obtained in Art. 193, on the supposition that the motion

took place *in vacuo*.   In practice, the resistance of the air must
be allowed for.   How this is done is explained by Newton in
a scholium at the end of his section on the Laws of Motion.
The effect of the air is, however, very slight, except when the
balls are very light.

Let both balls be drawn aside to points G and H of equal
height, and simultaneously let go, they will meet at their
lowest points.   Then by the preceding paragraph the velocity
of each at that point is known, and thus their relative velocity
at the moment of meeting is determined.

Also, if the heights of the points E and F, to which they
rise after impact, be observed, we shall know the velocity with
which each rebounded, and, therefore, their relative velocity
at the moment of separation.

The ratio e of the latter to the former relative velocity is
found to be pretty nearly constant, whatever be the weights
or sizes of A and B, or whatever be the height of H and G, as
long as the substances, of which A and B are formed, remain
the same.

The same result follows if keeping one ball at rest we draw
only the other aside.

198. We add some results, as to the value of e, taken from
Mr. Hodgkinson's paper :—

> Both balls made of cast iron, $e = .66$
> One cast iron, the other lead, $e = .13$
> One lead, the other boulder stone, $e = .17$
> Clay ball, soft brass ball, $e = .16$
> Both balls of glass, $e = .94$
> Both balls of ivory, $e = .81$
> Cork ball, ivory ball, $e = .60.$

Newton found that for 2 balls of wool bound up very com-
pactly the value of e was $\dfrac{5}{9}$.

EXAMPLES.—XXXIV.

(1.) If the length of a sling be 2 feet, with what velocity must a stone placed in it revolve in a vertical circle that it may be just prevented from falling out?

(2.) A particle is projected from the lowest point of a smooth circular hoop, along the circumference. Show that, if it passes through the centre, the velocity of projection was $\sqrt{ag}\sqrt{2+\sqrt{3}}$, a denoting the radius. Also find the arc described before leaving the hoop.

(3.) A particle is projected up the inside of a smooth vertical circle from the lowest point, with an initial velocity $=2\sqrt{ag}$; a being the radius of the circle. Find at what point it will leave the circle, and show that it will pass through a point in the vertical diameter at a height $\frac{25}{16}a$ above the lowest point. Find also the point at which it strikes the circle again, and its velocity at the moment of arriving there.

(4.) A particle whirled by a string in a vertical circle quits the circle, the string becoming slack; prove that the circle is the circle of curvature of the subsequent path, and that, if the particle quit the circle at a point 60° from the highest point, its subsequent path will meet the circle at the lowest point at an angle $\tan^{-1}3\sqrt{3}$, and the particle will then oscillate through an angle $2\cos^{-1}\frac{15}{16}$.

(5.) A heavy particle slides down a tangent to a vertical circle, starting from the point where this tangent meets the vertical diameter, and then slides on the circle; determine where it will leave the circle, and find the condition that it may just not slide upon it.

(6.) A heavy particle slides from the highest point of an ellipse, whose minor axis is vertical, down the arc; show that it will leave the curve at the end of the latus rectum, if the ratio of the minor to the major axis is given by the equation
$$x^3 - 4x + 2 = 0.$$

(7.) A heavy particle, resting at the vertex, which is also the highest point, of a smooth parabolic tube, is slightly displaced; show that, when it has attained a distance from the focus equal to the latus rectum, its pressure on the curve is $\frac{1}{8}$ of its weight.

(8.) **A** smooth circular cylinder of given height h, and whose radius is a, is **divided** into two equal parts by a plane through the axis, which is vertical, and one part **is** removed. A particle is projected horizontally **with** a given velocity from one end of the **upper** rim. Prove that the **greatest** possible velocity of projection, in order that the particle **may reach** the ground before it leaves the semi-cylinder, is $\pi a \sqrt{\dfrac{g}{2h}}$, and supposing the velocity greater than this, find the point where it will strike the ground.

(9.) A heavy particle is projected from the vertex of a parabolic **arc**, whose plane is vertical and axis horizontal, with such a velocity that after leaving the arc it describes an equal parabola. Show that it will **cross the axis** at a distance from the vertex $= \dfrac{3+2\sqrt{3}}{4}$. the latus rectum.

(10.) A smooth heavy particle is projected from the lowest point of a fixed circular arc, whose plane is vertical, up the curve with a velocity due to the diameter. Find the range of the particle on the horizontal plane through the point of projection.

(11.) A heavy **particle is projected from** a given point with a given velocity inside a parabolic tube whose axis is vertical ; show **that on escaping from the tube** it will describe a parabola whose focus for different lengths of the tube lies in a certain circle.

(12.) Two beads of equal weight are sliding down a perfectly smooth circular wire in a vertical plane, and are, at the same instant, at the **ends** of a vertical chord subtending a right angle **at** the centre ; find the velocity and direction of motion of their centre of gravity at that instant, each bead having **been** set moving from the highest point **with an** indefinitely small velocity.

(13.) A heavy particle slides in a smooth parabolic tube, whose **axis** is vertical and vertex downwards and latus rectum $= 4a$, starting from a point at a distance d from the focus. Prove that the pressure on the tube, **when** the particle is at a distance r from the focus, is $d\sqrt{\dfrac{a}{r^3}}$. weight of particle.

(14.) A particle is suspended **from** a fixed point by an inextensible string ; find the velocity with which it must be projected when at its lowest point, so that its path after the string has ceased to be stretched **may pass** through the point of suspension.

(15.) **From** a point upon **the** inside of the surface of a smooth cir-**cular hollow** cylinder, and inside, a particle is projected in a direction **making** an angle *a* with **the** generating line through the point ; find the velocity of projection that the particle may rise to a given height *h,* and the condition that the highest point may be vertically above the **point** of projection.

(16.) **A number of equal circular arcs of the same** curvature, and smooth, **are so united in order as to form a** continuous undulating plane curve by means of reversing every alternate one ; and the length of the arcs is such that the height of each wave is one-sixth of its ex-treme width. The plane of the curve being vertical and the summits of the waves in a horizontal line, a particle is projected horizontally with the greatest possible velocity from the bottom of a wave so as to **run** along the curve without **ever flying** off. Show that the velocity at the summit of a wave $= \dfrac{1}{\sqrt{2}}$ the **velocity at the bottom.**

199. *Motion in an inverted cycloid having its axis vertical.*

As in Art. 193, we will suppose that the particle is a bead, of **mass** *m*, sliding on a wire, or in a tube, which forms the cycloid.

Let *AB* be **the vertical axis,** *DBE* the base of the cycloid, *C* the centre, and *a* **the radius of the auxiliary circle.** See Chap. **VI.**

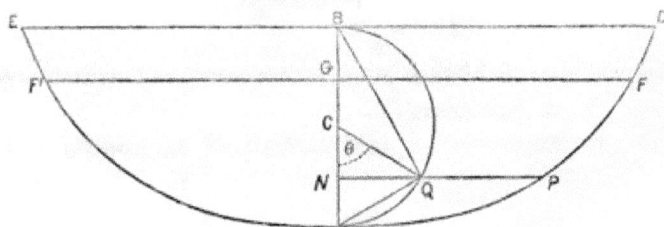

FIG. 57.

Let *v* be the velocity of the particle when at any point *P*.

Draw *PQN* perpendicular to *AB*, cutting the auxiliary in *Q* and *AB* in *N*. Let $ACQ = \theta$. Join *BQ, AQ.* Then the straight line BQ is parallel to the normal at P to the cycloid, and the radius of curvature at $P=2$ the chord $BQ=4a\cos\frac{\theta}{2}$.

Also, the straight line AQ is parallel to the tangent to the cycloid at P, and the arc AP (or s) $=2$ the chord AQ $=4a\sin\frac{\theta}{2}$.

Now the forces acting on the particle are,

(1.) Its weight, mg, acting vertically downwards, *i.e.* parallel to BA;

(2.) The pressure, R, of the curve acting along the normal at P, *i.e.* parallel to QB.

So that the sum of the resolved parts in the direction of the normal $=R-mg\cos\frac{\theta}{2}$.

And the sum of the resolved parts in the direction of the tangent $=mg\sin\frac{\theta}{2}$.

Hence, by Art. **175**, we have

$$\frac{mv^2}{4a\cos\frac{\theta}{2}}=R-mg\cos\frac{\theta}{2};$$

$$\therefore R=\frac{mv^2}{4a\cos\frac{\theta}{2}}+mg\cos\frac{\theta}{2}.$$

We may remark that R is never negative, and therefore the pressure never acts outwards.

Also, the acceleration in the direction of the tangent at P

$$=\frac{mg\sin\frac{\theta}{2}}{m}=\frac{g}{4a}\cdot s. \quad \text{Comp. Newton, Prop. X.}$$

As a particular case, suppose the particle to start from rest at D, then, Art. **187**, $v^2=2g.BN$

$$=2g(a+a\cos\theta)=4ag\cos^2\frac{\theta}{2};$$

$$R = \frac{4mag \cos^2\frac{\theta}{2}}{4a \cos\frac{\theta}{2}} + mg \cos\frac{\theta}{2}$$

$$= 2mg \cos\frac{\theta}{2},$$

which is double what R would be, if $v=0$, *i.e.* if the particle were placed at P instead of having moved to it from D.

200. Let the particle start from rest at any point F.

Draw FGF' perpendicular to AB, meeting AB in G and the cycloid again in F'.

Then the velocity of the particle on arriving at A will be $\sqrt{2g.AG}$; and passing through A it will move to F' and come to rest, for after leaving A its velocity gradually decreases, and at $F' = \sqrt{2.g.AG - 2g.AG} = 0$.

Immediately after coming to rest at F' it begins to descend, and, passing through A with the same velocity, $\sqrt{2.g.AG}$, as before, ascends to F and then returns. Thus it continues to oscillate between F and F'.

Now a curious property of this curve, and the one that makes the motion in it remarkable, is that wherever in AB the point F may be, at which it starts from rest, *the time of a complete oscillation*, from F to F' and back to F, is always the same. It is on this account that the cycloid is said to be an *isochronous* curve.

201. PROP. *To prove that an inverted cycloid with its axis vertical is an isochronous curve, and to find the time of an oscillation.*

Let AB be the vertical axis, DBE the base.

On AB as diameter describe the auxiliary circle AQB.

Let a particle start at a point F from rest.

Draw FGF' perpendicular to AB, meeting AB in G and the cycloid again in F'.

On AG as diameter describe a semicircle GpA, and let c be its centre.

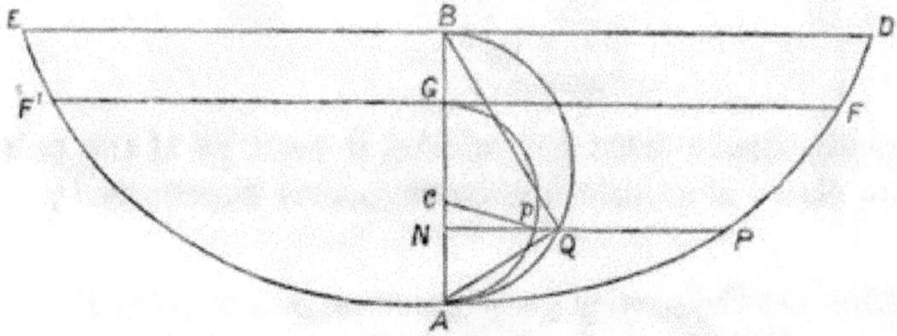

Fig. 58.

Let P be the position of the particle at any time.

Draw $PQpN$ perpendicular to AB, meeting the auxiliary circle in Q, the second circle in p, and AB in N.

Thus, as P descends from F to A, p moves round the semi-circle from G to A.

We will now determine the velocity of p.

Join BQ, AQ. We shall require the following geometrical relations :—

$AQ = \sqrt{AB.AN}$, *Euc.* VI. 8, Cor., BQA being a triangle right-angled at Q, and QN perpendicular to the hypothenuse ;

$pN = \sqrt{GN.AN}$, for similar reasons.

The point p and the particle P being always on the same horizontal line, their velocities in the vertical direction must always be equal ;

i.e. the vertical component of the velocity of $P =$ that of p.

Now the velocity of P is $\sqrt{2.g.GN}$; and the direction of its motion, being the tangent at P to the cycloid, is parallel to AQ, and therefore makes with the vertical an angle $(= BAQ)$,

whose cosine is $\dfrac{AQ}{AB} = \dfrac{\sqrt{AB.AN}}{AB} = \sqrt{\dfrac{AN}{AB}}$.

Hence the vertical component of the velocity of P

$$= \sqrt{\frac{2g.GN.AN}{AB}} = \sqrt{\frac{2g}{AB}} \cdot pN.$$

Again, let v denote the velocity of p. The direction of its motion is the tangent at p to the small circle, and therefore makes with the vertical the same angle that the normal cp at p makes with the horizon, *i.e.* the angle cpN, whose cosine is $\frac{pN}{cp}$.

Hence the vertical component of v is $\frac{v.pN}{cp}$.

Therefore $\sqrt{\frac{2g}{AB}} \cdot pN = \frac{v.pN}{cp}$; $\therefore v = \sqrt{\frac{2g}{AB}} \cdot cp$;

i.e. the velocity of p is uniform.

Again, since the time taken by P to descend from F to A is the same as the time taken by p to move round the semi-circle GpA, whose length is $\pi.cp$,

$$\therefore \text{ this time} = \frac{\pi.cp}{\sqrt{\frac{2g}{AB} \cdot cp}} = \pi \sqrt{\frac{AB}{2g}}.$$

Now the particle will evidently take the same time to move from A to F' as from F to A. Therefore the time from

F to $F' = 2\pi \sqrt{\frac{AB}{2g}}$.

And the time from F' back to F will be the same as from F to F'.

Hence the complete time (T) of oscillation $= 4\pi \sqrt{\frac{AB}{2g}}$.

Now $AB = 2a$; $\therefore T = 4\pi \sqrt{\frac{a}{g}}$.

Hence the time of oscillation in a cycloid is the same whatever the amplitude.

202. Prop. *To make a particle P, hung from a fixed point O by a string OP, move in a cycloid.*

Let *OA* be the position of *OP* when vertical.

Bisect *OA* in *B*.   Draw *DBE* horizontally through *B*.

Describe the cycloid *EAD* with its vertex at *A* and axis equal to *AB*.

Then *EAD* is the cycloid in which *P* can be made to oscillate.

Draw *DOE* the evolute of the cycloid, Art. **83**; then, if *DOE* is a material curve, when we move *P* from *A* towards *D*, keeping the string stretched, the latter will wrap round *OD*, and *P* will trace out the semi-cycloid *AD*.

Now, if *P* is held in any position *F* on *AD* and then let go, the string will unwrap; and, always being a tangent to *OD*

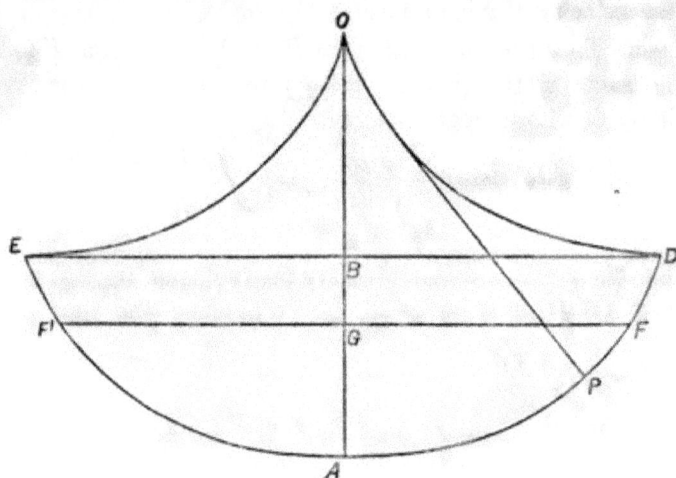

Fig. 59.

where it leaves *OD*, it will everywhere be perpendicular to the direction of the motion of *P*, and will therefore, by its tension, be capable of exerting the same force on *P*, as *AD* would by its pressure, if it were a material curve and *P* sliding on it as in Art. **201.**

Hence P will move exactly as if it were sliding on a material curve DAE.

On arriving at A, P ascends AE, and the string wraps on to OE till P comes to F'' on a level with F. After this it retraces its path back to F.

The whole oscillation is performed in the time $4\pi\sqrt{\dfrac{AB}{2g}}$.

Let l denote the length of the string, then $AB = \dfrac{l}{2}$;

$\therefore$ the time of a complete oscillation $= 4\pi\sqrt{\dfrac{l}{4g}} = 2\pi\sqrt{\dfrac{l}{g}}$.

203. *Simple Pendulum.*

DEF. A particle hung by a fine thread from a fixed point, and oscillating in a circle, forms with the string what is called a *simple pendulum.*

PROP. *To find the time of oscillation of a simple pendulum.*

For a short space on each side of A (Art. 202) the circle whose centre is O, being the circle of curvature of the cycloid at A, may be regarded as coinciding with the cycloid; and for positions of P within these spaces the bending of the string by OD or OE may be neglected; so that the motion of P within these spaces is the same, whether it be considered as moving on the cycloid, as in Art. 202, or on the circle whose centre is O and radius OA.

Hence the time of a complete small oscillation of a simple pendulum $= 2\pi\sqrt{\dfrac{l}{g}}$, where l is the length of the pendulum.

Thus, if we shorten the string, we shorten the time of oscillation; and if we shorten the string to one quarter of its length, we shorten the time of oscillation to one-half its former length.

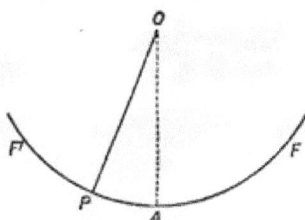

Fig. 60.

Obs. If we have a pendulum OP oscillating between F and F', the motion from F to F' and back again to F is what we have called the complete oscillation. Often, however, the motion from F to F' is spoken of as a single oscillation. In this case the time of oscillation

$$=\pi\sqrt{\frac{l}{g}}.$$

By a seconds pendulum we mean one which makes one oscillation (as from F to F'') in a second, and one such oscillation is called a beat.

204. From the formula

$$T=2\pi\sqrt{\frac{l}{g}}, \quad . \quad . \quad . \quad . \quad (1)$$

we can determine the value of g at any place.

If we set a pendulum swinging, and observe the time (T) of each oscillation in seconds, and measure the length (l) of the pendulum in feet, we can obtain the value of g from (1).

For instance, it is found that the length of the seconds pendulum in London is 39·1393 inches nearly.

Here then $T=2$ and $l=\dfrac{39\cdot1393}{12}$.

Hence $g=\pi^2\left(\dfrac{39\cdot1393}{12}\right)=32\cdot216$ approximately.

205. Let h be the height of a mountain, r the radius of the earth, g the acceleration of a falling body near the base of the mountain. Then, since the force of gravity varies inversely as the square of the distance from the centre of the earth, the acceleration of a falling body at the summit is $g.\left|\dfrac{r}{r+h}\right|^2$, roughly speaking.

Then, if T, T' be the times of oscillation of a pendulum of length l at the base and summit, $T=2\pi\sqrt{\dfrac{l}{g}}$,

$$T'=2\pi\sqrt{\frac{l}{g}.\frac{r+h}{r}};$$

$$\therefore T'=T\left(\frac{r+h}{r}\right).$$

Ex. A pendulum beats seconds at the foot of a mountain two miles high. How many beats will it make in an hour at the summit ?

We may take the radius of the earth to be 4000 miles; therefore $\dfrac{h}{r} = \dfrac{1}{2000}$. Also $T = 2$. Therefore $T' = 2\left(1 + \dfrac{1}{2000}\right)$.

Therefore, when it is at the summit, the time of each beat, being half the complete oscillation, $= \left(1 + \dfrac{1}{2000}\right)$ seconds.

Hence in one second the pendulum makes $\dfrac{2000}{2001}$ beats;

$\therefore$ „ hour „ $3600 \times \dfrac{2000}{2001}$ beats.

Further, in one hour it makes at the summit $3600\left(1 - \dfrac{1}{2001}\right)$ beats, and therefore it loses $\dfrac{3600}{2001}$ beats by being carried up the mountain. This, by continued fractions, can be shown to be about nine beats in five hours.

Conversely, if we obtained the times (T and T') of oscillation of the same pendulum at the foot and summit, we could determine the height of a mountain.

EXAMPLES.—XXXV.

(1.) When a particle starts from the cusp of an inverted cycloid, the vertical velocity of the particle is greatest when it has completed half its vertical descent.

(2.) If a particle start from a cusp of an inverted cycloid, the pressure on the curve at any point is proportional to the radius of curvature, and the part of it due to gravity is equal to the part due to centrifugal force.

(3.) If a particle start from one end of the arc of an inverted cycloid, show that, when it has fallen through half the distance, measured

N

along the arc, to the lowest point, it will have accomplished three-fourths of its vertical descent, and that two-thirds of the time of descent will have elapsed.

(4.) The time in which a particle can slide from the highest point, C, of an inverted cycloid to the lowest, A, is divided into two equal parts at the point P of the curve. Show that the arc AP : arc AC = velocity acquired at A in an oscillation commencing at P : velocity acquired at A when motion begins at C.

(5.) Show that the time of descent from any point on a cycloid to the corresponding point on the evolute is the same.

(6.) A particle, P, slides down a tube in the form of an inverted cycloid. A string being attached to it passes out of the tube at its highest point, and carries a weight, W, less than that of P, which descends vertically. Find the position beyond which P must not be placed, so that it may not be drawn out of the tube; and determine the time of oscillation.

(7.) A circular fine smooth tube contains a particle which makes small oscillations in it, while it is with its plane inclined at an angle θ to the horizon. Show that the period of these oscillations is to the period in the vertical plane as $1 : \sqrt{\sin\theta}$.

(8.) A tube in the shape of a cycloid is placed with its axis vertical and vertex downwards; one of two particles whose elasticity (e) is equal to the ratio of their weights, is placed at the vertex, and the other let drop into one end of the tube. Show that, if v be the velocity (due to the height of the cycloid) with which it impinges on the first, the velocity of the first after the nth impact will be, according as it is the lighter, or the heavier, of the two

$$\frac{1-(-e)^n}{1+e}v, \text{ or } \frac{1-(-e)^n}{1+e}ev.$$

(9.) Find the correction to be made in the pendulum of a clock which loses $10''$ per hour.

(10.) A pendulum, which beats seconds at the foot of a mountain, loses $10''$ a day when taken to its summit; find the height of the mountain, assuming the earth's radius 4000 miles, and neglecting the attraction of the mountain.

(11.) In marching in slow time, the length of a pace is 30 inches, and the time is given by the semi-oscillation of a plummet 24·96 inches long; find the rate of marching.

(12.) If a clock at a place A on the earth's surface keeps true time, and when taken to another place B loses a minute daily, but is made again correct by shortening the pendulum by $\frac{1}{n}th$ part of an inch, find the accelerating effect of gravity at A.

(13.) A body starts from rest at the highest point of a cycloid, whose axis is vertical and vertex downwards, and slides down the arc, and when it has described half the distance to the vertex, it is allowed to move freely; show that it will reach the tangent at the vertex at a point whose distance from the vertex is equal to

$$\left(\frac{\pi}{3} + \frac{3}{3\sqrt{7} + 5\sqrt{3}}\right) \cdot \text{ the radius } (a) \text{ of the generating circle.}$$

Show also that it will reach the tangent to the vertex earlier than it otherwise would have done by an interval equal to

$$\sqrt{\frac{a}{g}}\left(\frac{\pi}{3} - \frac{2}{\sqrt{7} + \sqrt{3}}\right).$$

(14.) If a particle slide from the highest point down the concave arc of a smooth cycloid whose axis is vertical, prove that the ratio of the time of describing any arc to the time of falling vertically through the diameter of the generating circle is half the circular measure of the angle through which the generating circle must have turned to describe the arc.

(15.) Prove that if a heavy particle slide freely from the highest point of a cycloid, of which the axis is vertical and vertex downwards, the angular velocity of the generating circle passing through the point will be constant, and inversely proportional to the square root of its radius.

MISCELLANEOUS EXAMPLES.

(1.) At the same moment one body is dropped from a given height, and another is projected vertically upwards from the ground with just sufficient velocity to attain that height; compare the time in which they will meet with the time in which the first would have fallen to the ground.

(2.) A particle of weight equal to 10 ounces is suspended from a fixed point by a string l feet long. The particle describes a horizontal circle with an uniform velocity of 50 yards a minute; find the tension which the string must be capable of bearing so as just not to break.

(3.) Two particles A and B, of masses $8m$ and m respectively, lie together on a smooth horizontal plane, connected by a string which lies loose on the plane; B is projected at an elevation of 30° with a velocity equal to g. If the string becomes tight the instant before B meets the plane again, and breaks when it has produced half the impulse it would have produced if it had not broken, and if the particle rebounds at an elevation of 30°, show that the elasticity of B is $\dfrac{5}{9}$.

(4.) A hollow spherical shell has a small hole at its lowest point, and any number of particles start down chords from the interior surface at the same instant, and pass through the hole and then move freely. Show that before and after passing through the hole they lie on a spherical surface; and determine its radius and position at any instant.

(5.) Two perfectly elastic balls are dropped from different heights upon a horizontal plane; find the condition that the centre of gravity of the two may again rise to its original height.

(6.) A particle is projected horizontally from the lowest point of a smooth fixed sphere, and inside it, so as to pass through the centre of

196

the sphere. Show that its first impact against the sphere takes place at an angular distance $\sin^{-1}\dfrac{\sqrt{2}}{3\sqrt{3}}$ from the point of projection.

(7.) A number of heavy particles are projected simultaneously from a point; if tangents be drawn to their paths from any point in the vertical line through the point of projection, prove that the points of contact will be simultaneous positions of the particles.

(8.) A parabola is placed in a vertical plane, and its axis is inclined to the vertical, S is the focus, A the vertex, and Q the point in the curve which is vertically below S. If SP be the line of quickest descent from the focus to the curve, show that ASP is equal to twice the angle PSQ.

(9.) If the point and velocity of projection of a particle be given, prove that the locus of the focus of the path is a circle, and thence show that the locus of the vertex is an ellipse.

(10.) The vertical component velocity of a body is $\frac{4}{5}$ of the whole; find the horizontal velocity and the direction of motion.

(11.) A man stands on a horse which canters round a circus with uniform velocity. State the forces acting on him, and the condition of relative equilibrium. If he spring up, how must he do so in order to light upon the back again?

(12.) Two smooth planes, OA, OB, each inclined to the horizon at the same angle a, which is less than $\dfrac{\pi}{4}$, intersect in a horizontal line.

An inelastic particle descends from rest at A, show that the time which elapses before it is reduced to rest is to the time of descending AO as $\cot^2 a : 1$.

(13.) If the unit of space be a fathom, of time half a minute, and of weight a pound avoirdupois, compare the new units of velocity, acceleration, and density with the ordinary units.

(14.) If the area of a field of 10 acres be represented by 100, and the acceleration of a falling weight by $58\frac{2}{3}$, find the unit of time.

(15.) Show how the depth of a mine may be ascertained by means of a pendulum, supposing the force of gravity within the earth to vary directly as the distance from the centre.

(16.) Two bodies are dropped from points P and Q, on a smooth inclined straight line XY, and reach the ground with the same velocity; prove that PQ is perpendicular to XY.

(17.) Heavy particles slide from rest down chords of a vertical circle passing through the highest point. Prove that the locus of the foci of the parabolic paths they describe after leaving the chords is a circle, and that of their vertices an ellipse whose axes are in the ratio 2 : 1.

(18.) A ball A impinges obliquely on a ball B at rest, their mutual elasticity being e. Show that the maximum deviation of A is

$$= \tan^{-1}\frac{(1+e)B}{2\sqrt{(A+B)(A-eB)}} \; ;$$

provided $A > eB$; and examine the case when $A < eB$.

(19.) Two inclined planes (a, β) are set back to back, and two weights, P, Q, are placed on them, being connected by a stretched string passing over the common vertex of the plane. If P predominates, and the motion takes place in a vertical plane, find the acceleration of P and Q, the tension of the string, and the pressures on the planes and the ridge. If the planes rest on a rough horizontal table, show that it will slip along it if the coefficient of friction be

$$< \frac{(P\sin a \sim Q\sin\beta)(P\cos a + Q\cos\beta)}{(P+Q)^2 - (P\sin a - Q\sin\beta)^2}.$$

(20.) The point O is the orthocentre of a triangle ABC, and the component velocities of a point are represented by OA, OB, OC; prove that the resultant velocity is represented by twice the distance between O and the centre of the circumscribing circle.

(21.) Two particles are let fall at different instants down two inclined planes from the same point in the horizontal ridge in which they meet; prove that each describes relatively to the other a straight line with uniform acceleration.

If the direction of the relative motion make an angle θ with the plane which bisects the external angle between the given planes, then $2\theta = a \sim \beta$, where a and β are the inclinations of the given inclined planes to the horizon.

(22.) A fine smooth inextensible string is placed on two inclined planes joined at the highest point and resting on an horizontal table, and is tacked at one end to one of them, the inclinations of the planes to the horizon being a, β, and the lengths of the string on each x, y, respectively. If the tack be suddenly removed, find the alteration in the finite pressure on the vertex, and show that the pressure on the table is instantaneously diminished by $\left(\dfrac{x\sin a - y\sin\beta}{x+y}\right)^2 \times$ (weight of the string).

(23.) **Two** equal balls of mutual elasticity e are simultaneously projected from two adjacent corners of a square table so as to meet in the middle; find the planes and the latera recta of their orbits after leaving the table.

(24.) If APB be a circle revolving uniformly about A, the end of the diameter AB, while P moves in any given manner in the circle. Find the acceleration of P in the direction of the tangent to the circle, and perpendicular to it.

(25.) OA, OB are two lines of railway crossing each other at right angles. A train is travelling with a given velocity from A towards O, containing troops that have guns whose maximum range is c, and when its distance from O is a (a being $> c$) a train of the enemy starts from O and travels along OB with a known velocity. Find the first and last times when the enemy's train can be shot.

(26.) If the perimeter of a circle be the unit of length, and the time of descent down the vertical diameter the unit of time, show that the accelerative effect of gravity is measured by $\dfrac{2}{\pi}$.

(27.) **Particles** are projected down the chords of a vertical circle drawn through the highest point with velocities proportional to the chords; show that they always lie on a circle.

(28.) If a be the distance between two points at any time, V their relative velocity, and u, v the resolved parts of V in and perpendicular to the direction of a, show that their distance when they are nearest to each other is $\dfrac{av}{V}$, and that the time of arriving at this nearest distance is $\dfrac{au}{V^2}$.

(29.) A train is moving at the rate of 30 miles an hour on smooth horizontal rails, when the guard observes that a bridge at the distance of 274 feet from the train has fallen in, he immediately puts on the break, which is equivalent to suddenly attaching to the train a weight of 10 tons hanging freely by means of a string passing over a smooth pulley; given that the weight of the train is 100 tons, show that it is a matter of some consequence to the passengers whether g equals 32 or 32·2.

(30.) If the unit of space be 8 feet, and the acceleration of a body falling under the action of gravity be 4, find the numerical value of the moon's mean velocity, its periodic time being $27\frac{1}{2}$ days, and mean distance 240,000 miles.

(31.) A smooth ball moving on a horizontal plane impinges on a ball of $\left(\dfrac{1}{n}\right)th$ its mass at rest on the plane; the line joining the centres of the balls being inclined at an angle a to the horizon at the instant of impact, and the whole motion taking place in one plane. Prove that the first ball will rise vertically from the impact provided that the coefficient of elasticity between the balls be $n + \tan^2 a$.

(32.) A number of particles of different elasticities slide down a smooth inclined plane, through the same vertical height, and impinge on a horizontal plane at its foot; show that they all, throughout their subsequent motions, describe portions of the same parabola.

(33.) From a point A a ball X is projected vertically upwards, and at the same moment an equally heavy ball Y is projected from a point B, so as to strike X horizontally when X is at the highest point of its path; AB is a line on level ground, and perpendicular to BA produced is a perfectly elastic vertical screen at a distance $\frac{1}{2} AB$ from A. Show that the balls strike the ground at the same point.

(34.) How would the oscillation of a cycloidal pendulum be affected by the pendulum being carried—1st, uniformly; 2dly, with uniform acceleration, along a straight railway?

(35.) Two buckets of given weights are suspended by a fine string placed over a fixed pulley; at the centre of the base of one of the buckets a frog of given weight is sitting; at an instant of instantaneous rest of the bucket, the frog leaps vertically upwards so as just to arrive at the level of the rim of the bucket. Find the ratio of the absolute length of the frog's vertical ascent in space to the length of its bucket, and prove that the time which elapses before the frog again arrives at the base of its bucket is independent of the frog's weight.

(36.) If the area of a circle, of radius r feet, be the unit of area, and the velocity, which a heavy particle would acquire in falling freely through a space equal to the diameter, the unit of velocity; find the acceleration with which a point would describe the circle uniformly in t seconds.

(37.) Three equal uniform elastic spheres, A, B, C, are placed on a horizontal table, so that the centres of A, B, C are in a straight line, the centre of the middle one, B, being at equal distances from the centres of A and C. Another equal sphere, D, is projected along the table, so as to strike B; B then proceeds to strike one of the outer of the three balls and D the other; prove that the radius of any one of the spheres has to the distance between A and C a greater ratio than 1 to $4\sqrt{2}$.

(38.) A particle is projected with a velocity of 200 yards a second, and after passing through 6 inches of a substance which retards it with a uniform force, it is found to have lost two-thirds of its velocity. Find the time taken to pass through.

(39.) A vertical wheel moves on a perfectly rough horizontal plane with the velocity it would acquire by falling through a height equal to half its radius ; a particle flies off at a point P. Show that the focus of the parabola described by the particle is the foot of the perpendicular from the lowest point of the wheel upon the radius through P ; and that the focal distance of P is a mean proportional between the semi-latus-rectum and the radius of the wheel.

(40.) Two particles are simultaneously projected with velocities v_1, v_2, and directions a_1, a_2 with the vertical. Prove that after an interval

$$\frac{v_1 v_2 \sin(a_2 - a_1)}{g(v_2 \sin a_2 - v_1 \sin a_1)}$$

their directions are parallel, and find their relative velocity.

ANSWERS.

I. p. 3.

(1.) 400 yards.   (2.) 9000.   (3.) 176 feet ; $19\frac{1}{3}$.

(4.) $2\frac{1}{22}$ miles an hour.   (5.) 352.   (6.) $\frac{5}{44}$ minutes ; $3\frac{9}{22}$.   (7.) 1408.

II. p. 4.

(2.) 176.   (3.) 396″ ; 1320′.   (4.) 154 : 40.   (5.) $\frac{9}{4200}$.   (6.) 440 : 3.

(7.) 11 : 9.   (8.) 5 : 56.   (9.) 88 : 3.   (10.) $\frac{9}{154}$.   (11.) $\frac{44}{45}$.

(12.) 10 miles an hour ; 6 miles an hour.

III. p. 7.

(1.) 70.   (2.) 15.   (3.) 13.   (4.) $\frac{352}{15}$; 352.   (5.) $3333\dot{\cdot}3$.   (6.) 126·72.

(7.) 350 yards.   (8.) The measure is $\frac{88}{35}$.   (9.) $\frac{7}{240}$.   (10.) 1025 feet.

(11.) $50\sqrt{8101}$ feet ; $100\sqrt{13,690}$ feet.   (12.) $400\sqrt{7}$ yards.
(13.) The unit velocity.

IV. p. 8.

(1.) 18 inches.   (2.) $11\frac{2}{3}$ inches.   (3.) $-\frac{1}{22}$.

(4.) By a line 12·5 inches long, making 45° with the first.

V. p. 11.

(1.) 16,800.   (2.) $3''\frac{6}{7}$.   (3.) 4 inches.   (4.) $\frac{12}{29}$ seconds.   (5.) $4''\frac{2}{3}$.

(7.) $\frac{1}{88}$ minutes.   (8.) $\frac{1}{1760}$ minutes.   (10). $26\frac{7}{25}$ inches.

VI. p. 15.

(1.) A velocity $3\sqrt{10}$ feet a second, making $\tan^{-1}\frac{1}{3}$ with the second.

(2.) $3\sqrt{3}$; 3.   (3.) $5\sqrt{17+4\sqrt{2}}$ miles an hour at $\tan^{-1}\frac{2\sqrt{3}}{1+2\sqrt{2}}$.

(4.) $\tan^{-1}\frac{2}{3}$ with the 3 ; By $\tan^{-1}\frac{5}{12}$.   (6.) $\sin^{-1}\left(\frac{v_2\sin\beta - v_1\sin\alpha}{r_1+r_2}t\right)$. •

VII. p. 19.

(1.) $v\sqrt{2}$ in direction SE.   (2.) $2\sqrt{13+6\sqrt{2}}$ miles an hour in direction at $\tan^{-1}\frac{3\sqrt{2}-2}{7}$ with S and N line.

(3.) $2\sin 15°$. (initial velocity), at $105°$ to its initial direction.

(4.) $\sqrt{31}$ miles an hour, at $\tan^{-1}\frac{3\sqrt{3}}{8}$ to its initial direction.

VIII. p. 23.

(1.) 1200 units.   (2.) 27,000 feet per second.   (3.) ·2 feet per second.
(4.) 10,800.   (5.) One second.   (6.) ·5 feet per second.

IX. p. 24.

(1.) 15 : 1.   (3.) 120.   (5.) 2 : 3.   (6.) 140.   (7.) 20.   (8.) ·04.

X. p. 26.

(1.) 70.   (2.) 3.   (3.) 3.   (4.) 660.   (5.) 15.   (6.) 8400.

(7.) 10,000 feet per second.   (8.) $2\frac{2}{3}$.   (9.) 350 feet a minute.

(10.) 269,965 feet a minute ; 539,951 feet a minute.

(11.) 9600.   (12.) $\frac{7}{7200}$.

XI. p. 29.

(1.) 365; −355.

(2.) 8 inches.

(3.) $-\dfrac{75}{44}$. (4.) By a line $843\dfrac{3}{4}$ inches long, at 45° to the first.

XII. p. 31.

(1.) 96.   (2.) $\dfrac{1}{165}$.   (3.) $8\dfrac{1}{3}$ feet.   (4.) 1 : 4.

(7.) $\dfrac{a^2}{b}$ feet ; $\dfrac{at}{b}$ second.   (9.) One foot ; 2 inches.

(10.) $\dfrac{20}{161}$ feet ; $\dfrac{10}{161}$ seconds.

XIII. p. 35.

(1.) 5 at $\tan^{-1}\dfrac{4}{3}$ to the 3.   (2.) $300\sqrt{1753}$, at $\tan^{-1}\dfrac{32}{27}$ to the E line.

(3.) $30\sqrt{5+2\sqrt{3}}$, at $\tan^{-1}\dfrac{1}{4+\sqrt{3}}$ to the direction of initial motion.

(4.) $\dfrac{1}{\sqrt{2}}$, in direction bisecting the exterior angle between the initial and final directions of motion.

XIV. p. 40.

(1.) 322 ; $2\sqrt{966}$; 1610 ; 4830 ; $\dfrac{2530}{161}$.

(2.) $6g$.   (3.) 550.   (4.) 3 ; 22·5 feet from starting point.

(5.) $\dfrac{250}{161}$ seconds.   (6.) 65.   (7.) 8 feet per second per second.

(8.) 180 feet from starting point.   (9.) $\dfrac{20}{3}$; 20.

(10.) $\dfrac{15125}{322}$ feet ; $\sqrt{4313}$ feet a second.   (11.) $10\dfrac{\sqrt{3}-\sqrt{2}}{\sqrt{161}}$ seconds.

(12.) $\sqrt{gh}$.   (13.) $\dfrac{983,000 \mp 100,000\sqrt{2983}}{161}$ feet.

(19.) $\dfrac{a(2n-1)}{2a}$ feet.   (20.) $S = mvt + mn\dfrac{at^2}{2}$.   (21.) 5 yards.

XIV. p. 40—*continued.*

(23.) 44 feet; $\dfrac{1}{2}$.

(24.) $\dfrac{240}{11}$.

(25.) $(n \pm \sqrt{n^2 - 2n})$ seconds ; $g\sqrt{n^2 - 2n}$.

(26.) 1 foot, 1″.

(28.) $\dfrac{1}{32}$ seconds ; $\dfrac{1}{32}$.

(29.) $\dfrac{ft}{b}$ seconds, and $\dfrac{2af}{b^2}$ feet.

(30). 5.

XV. p. 50.

(1.) $\dfrac{1}{3520n^2}$ of an hour. (2.) If v, ω be linear and angular velocities of moving point, r distance from fixed point, p the perpendicular from fixed point on line of motion, $vp = \omega r^2$.

(3.) $v.AC.BC = \omega CP^2.CD$; $\dfrac{\omega^2 CP^4}{CD.AC.BC}$.

(4.) $\dfrac{v}{\text{diam.}}$.

(5.) If ω is angular velocity about focus,

$$\omega SP \sqrt{\dfrac{SP - AS}{SP + 3AS}}\; ;\; \omega^2 \dfrac{SP^{\frac{3}{2}}}{2\sqrt{AS}}.$$

(6.) If v is velocity of moving point, θ its angular distance from the given point, $v \cos\dfrac{\theta}{2}$; $\dfrac{v^2}{a} \sin\dfrac{\theta}{2}$, a being radius.

XVI. p. 55.

(1.) If v is velocity of hour hand, θ the angle through which it has gone since noon, the relative velocity is, $-12v \sin 12\theta + v \sin\theta$ parallel, and $12v \cos 12\theta - v \cos\theta$ perpendicular, to the noon line.

(4.) If a, β are the accelerations, θ the angle between their lines of motion, u, v their velocities at some given instant, the relative velocity is least at an instant whose distance from given one is

$$\dfrac{(av + \beta u) \cos\theta - ua - v\beta}{a^2 + \beta^2 - 2a\beta \cos\theta}.$$

(6.) Velocity of A with respect to B is $a\omega$, parallel to C.G., a being a side, ω the angular velocity of A about G the C.G.

(7.) $v(1 - e^2)$ at an angle to CA equal to BCP.

XVII. p. 66.

(1.) $289\cdot8$; $12\sqrt{415}$. (2.) 5. (3.) 3. (4.) 25. (5.) 49. (6.) 0.
(7.) $18-32\sqrt{2}$; $36+32\sqrt{2}$. (8.) 5430. (9.) 300. (10.) 1800.

XVIII. p. 78.

(1.) 90. (2.) 1 : 2. (3.) 5 : 3. (4.) 3 : 7 ; 3 : 7. (5.) 45.
(6.) Sufficient to generate the momentum $6\sqrt{3}$. (7.) $\sqrt{3}:1$.
(8.) $\sqrt{b^2+2gh}$.

XIX. p. 88.

(1.) 20 ; 10. (3.) 7 : 15 ; 7 : 15. (4.) 33 : 25. (5.) 7 : 66 ; 7 : 660.
(6.) 2 : 1 ; 1 : 2. (7.) Gravity.

XX. p. 93.

(1.) $\dfrac{320}{161}$. (2.) $112\cdot7$. (3.) 6. (4.) $\cdot75$. (5.) 60. (6.) $\cdot5$.

(7.) $19\cdot32$. (8.) $6\cdot25$. (9.) $\dfrac{125}{161}$. (10.) 5. (11.) $\dfrac{280}{161}$.

(12.) $150\sqrt{\dfrac{10}{161}}$. (13.) $100\sqrt{\dfrac{10}{161}}$. (14.) $10\sqrt{16\cdot1}$.

(15.) $3\dfrac{3}{160}$ inches. (16.) $\dfrac{24}{\sqrt{16\cdot1}}$ seconds. (17.) $\dfrac{100}{161}$ lbs.

(18.) 322. (19.) $15\dfrac{5}{161}$ lbs. (20.) $2\cdot0125$. (21.) $15\dfrac{85}{161}$.

(22.) 11 : 1890. (23.) If b'' is unit of time, a feet of space, $a=(32\cdot2)b^2$.

(24.) 160 lbs. (25.) 57 lbs., nearly. (26.) $31\dfrac{9}{161}$ oz.

(27.) $6\dfrac{144}{161}$ lbs., downwards ; $2\dfrac{68}{161}$ lbs., upwards. (28.) $\dfrac{25}{161}$ lbs.

XXII. p. 114.

(2.) $\dfrac{P^2}{Q^2-P^2}$. (certain height). (3.) 4 lbs.

(4.) $\dfrac{m\sin\alpha-m'\sin\beta}{m+m'}g$; $\dfrac{2mm'\sin\alpha\sin\beta}{m+m'}$. (5.) 6 on plane.

(6.) 12 on one side. (8.) $\dfrac{1}{4}$ second ; $\dfrac{9}{8}$.

XXII. p. 114—*continued*.

(10.) If $K = (P+Q)(p+q) + pq$, $\dfrac{P(3q-p) - pq - Q(p+q)}{K} g$, up,

$\dfrac{P(3p-q) - pq - Q(p+q)}{K} g$, up, $\dfrac{(P-Q)(p+q) - pq}{K} \cdot g$, down ;

$2P \dfrac{Q(p+q) + pq}{K}$, $4P \cdot \dfrac{pq}{K}$.

(11.) If m is mass of each body, c the length of each section, mrg,

$m\sqrt{2rcg}$.　　(12.) $2\frac{5}{8}$lbs., $1\frac{1}{8}$lbs.　　(13.) $\dfrac{3}{2}$. (given space).

(18.) $\operatorname{Sin}^{-1}\dfrac{20}{3g}$.　(22.) $\sqrt{gl}$ or $\sqrt{gh}$, according as $h >$ or $< l$. (23.) $\dfrac{22}{483}$.

(29.) 60° from highest point. (30.) If AB is vertical diameter when plane is vertical, AB' its position afterwards, then $AB'B = 90°$.

(31.) (1.) Makes equal angles with line and horizon ; (2.) Join point to highest point of circle ; (3.) Makes equal angles with given line and horizon, and goes through lowest point of circle.

(38.) $83 : 55$.　(40.) $(n+1)m : n(1+mn)$.　(43.) $4WW' : (W+W')^2$.

XXIII. p. 123.

(1.) Parabola, axis parallel to tube.　(2.) $\dfrac{1800}{g}$.　(3.) 25 yards.

XXIV. p. 127.

(1.) $g\sqrt{21}$, $\tan^{-1}3\sqrt{3}$.　　(2.) $\dfrac{5}{g}(\sqrt{6} - \sqrt{2})$.

(3.) $\dfrac{6g \sin 75° \pm \sqrt{9g^2(2+\sqrt{3}) - 240g}}{2g}$; $\dfrac{3}{2\sqrt{2}}\sqrt{9g^2 - 240g(2-\sqrt{3})}$.

(4.) $\dfrac{1000\sqrt{3} - 1}{3} \cdot \dfrac{1}{g}$; $\dfrac{100\,3\sqrt{2} - \sqrt{6}}{3} \cdot \dfrac{1}{g}$.　(5.) $\dfrac{400}{g}$.　(6.) $9\frac{3}{28}$ feet.

(7.) $\dfrac{2v^2}{g}$.　　(8.) 368 feet, nearly, taking $\sqrt{570} = 2$, $\sqrt{4830} = 69$.

(10.) 1946 nearly.　　(11.) 435 nearly.

(12.) $2h\{n \sin\theta \cos\theta \pm \cos\theta\sqrt{n + \sin^2\theta}\}$

(14.) $\tan^{-1}\dfrac{h_1 \sin e_1 \cos e_1 - h_2 \sin e_2 \cos e_2}{h_1 \cos^2 e_1 - h_2 \cos^2 e_2}$.

XXV. p. 131.

(18.) $t\sqrt{u^2 + v^2}$, where u, v are initial velocities.

XXVI. p. 137.

(1.) $7, 9\frac{1}{2}$. (2.) $B = 3A$. (4.) $8u$ and $2u$, in direction of original motion of $3m$. (10.) $10 : 6 : 3 : 1$.

XXVII. p. 140.

(1.) $\frac{1}{16}\sqrt{2459 + 280\sqrt{3}}$, $\frac{1}{16}\sqrt{4147 + 600\sqrt{3}}$, $\cot^{-1}\frac{5\sqrt{3} + 28}{40}$,

$$\cot^{-1}\frac{20 + 15\sqrt{3}}{32\sqrt{3}}.$$

(2.) $\frac{1}{3}\sqrt{59 - 12\sqrt{6}}$, $\frac{1}{18}\sqrt{3779 - 480\sqrt{6}}$, $\cot^{-1}\frac{3\sqrt{3} - 8\sqrt{2}}{9}$,

$$\cot^{-1}\frac{15\sqrt{3} - 16\sqrt{2}}{36\sqrt{2}}.$$

(3.) $\frac{3}{2}(\sqrt{6} - \sqrt{2})$, $\frac{3}{2}$, $\tan^{-1} - (\sqrt{3} + 2)$. (4.) 30°, $6\sqrt{3}$, $3\sqrt{3}$.

(9.) $V \sin^{-1}\frac{n-4}{2n}\pi$, each ball struck moves perpendicular to the line along which the striker then proceeds ; the rth ball struck starts with a velocity $V \sin^{r-1}\frac{n-4}{2n}\pi \cos\frac{n-4}{2n}\pi$.

(10.) $\frac{m}{\tan\alpha - 1}$, where m is the mass, α the angle **between the** previous direction of motion and the line of the blow.

(11.) **Describe** on AE, ED, two segments of circles containing angles equal to 135°. They intersect in F.

XXVIII. p. 143.

(1.) $\frac{5}{\sqrt{3}}$, at 30° to plane on other side of normal ; 90°. (2.) $\text{Tan}^{-1}\sqrt{e}$.

(5.) $2a = b$, or $3a = 2b$, etc. ; number of impacts is 3, or 7, etc., excluding those at point of departure.

(8.) In a direction at $\cot^{-1}\frac{e}{ex + y}\sqrt{c^2 - x^2 - y^2 + 2xy}$ to cushion, x, y, c being the 3 given distances.

XXIX. p. 149.

(2.) With notation of Art. 171, Ex. 1, the height h above centre is given by $ag(1+\cos\theta)=u^2\sin2\theta\{e(1-\cos\theta)^2-e^2(1-\cos\theta)\cos\theta-\cos\theta\}$, $u^2=2g(h+a\sin\theta)$.

(3.) 36 feet; $\sqrt{\dfrac{3}{5}}$. 　　(6.) $\dfrac{ev^2\sqrt{3}}{4g}$ from wall, $\dfrac{v\sqrt{3}}{g}$.

(7.) $e=1$, $1:r:r^2:$ etc., where $r=\dfrac{1-\sin\beta}{1+\sin\beta}$.

XXX. p. 150-155.

(1.) $\dfrac{1}{2}$. 　(6.) 100 **feet ; after** impact the highest descends **with a** velocity beginning at $5g-20$, and the lowest with **a** velocity $5g$.

(7.) 80, g being 32. 　(9.) Vertex of new path is on directrix of old, distance between their axes, being half the latus-rectum of either. Each ball reaches the ground at the instant the other would have **done,** if there had been no impact.

(11.) $\dfrac{2S}{abc}$. 　　(12.) Integer in $\dfrac{4na}{c}$. 　　(13.) $\operatorname{Tan}^{-1}\dfrac{5}{3}$.

(15.) $V^2=\dfrac{ag}{e^{2a+1}\sin2\theta}\cdot\dfrac{e^{2a+2}-1}{e-1}$, where a is distance between walls, and θ the angle of projection.

(18.) $\dfrac{2a}{u}\dfrac{1+e'}{1-e'-ee'-e}$, at distance $=\dfrac{2aee'}{1-e'-ee'-e}$.

(24.) $u\dfrac{m-2eM}{2M+m}$. 　　(27.) **End of latus-rectum.**

(29.) Height of directrix is raised by $\sqrt{\dfrac{3}{2}}$ of its previous height above point of impact, and latus-rectum remains unchanged in length.

(31.) $\sqrt{1-\dfrac{\sqrt{3}}{n}}$. 　　(36.) $\dfrac{h}{e^2\sin^3a}$.

XXXI. p. 158.

(2.) $\dfrac{1}{2}\sqrt{lg}$ horizontally **to right** and $\dfrac{1}{2}\sqrt{lg}$ vertically down.

(4.) C.G. has a velocity $\propto\sin2\beta$ perpendicular, and one $\propto\cos2\beta$ parallel to diagonal, β being one of the two angles into which the diagonal divides an angle of the rectangle.

o

XXXII. p. 164.

(1.) $f = a\omega^2$. See Art. 176, Ex. 2.

(4.) $\dfrac{2\pi}{\sqrt{\omega^2 - 18\cdot 2}}$ second, ω being the real angular velocity.

(5.) $4\sqrt{g}$.   (6.) $\sqrt{\dfrac{g}{a}\sec^2 30^\circ + \omega^2}$.   (8.) $\dfrac{27}{64}\dfrac{g}{\pi^2}$.

XXXIII. p. 168.

(1.) 3 ; 9 ; 6.   (2.) $\dfrac{1}{2}\dfrac{mm'}{m+m'}V^2$.   (3.) $\dfrac{1}{2}\dfrac{(m+m')m''V^2}{m+m'+m''}$.

(4.) 150 ; 15.   (5.) $\dfrac{W}{vt}$ (6.) Diminished by one half.   (7.) $\dfrac{\omega}{2\pi r}$.

XXXIV. p. 183.

(1.) $\sqrt{2g}$.

(3.) It leaves circle at a point at distance $\cos^{-1}\dfrac{2}{3}$ from highest point.

(5.) It leaves circle at a distance from highest point $= \cos^{-1}\dfrac{2}{3\cos\theta}$, θ being distance of point where it meets circle. It must meet circle at a distance from highest point $= \cos^{-1}\sqrt{\dfrac{2}{3}}$.

(8.) At a distance $\dfrac{v\sqrt{2gh} - g\pi a}{g}$ from edge of cylinder.

(10.) $a\sin\theta + a\sin 2\theta\{1 + \cos\theta + \sqrt{2 + 2\cos\theta + \cos\theta^2}\}$, a being radius, $a\theta$ the length of circular arc.

(12.) $2\sqrt{ag}$ at $\tan^{-1}(\sqrt{2}+1)$ to horizon.

(14.) $\sqrt{ag(2+\sqrt{3})}$, where a is length of string.

(15.) $\sqrt{2gh}\sec a$; $\pi a\cos a = h\sin a$, a being radius of cylinder.

XXXV. p. 193.

(6.) Distance of required, from lowest, point $= \dfrac{2W}{P}$ ·diameter of auxiliary circle ; $4\pi\sqrt{\dfrac{a(P+W)}{g.P}}$.

(9.) Shorten it by $\dfrac{719}{129,600}$ of its length. (10.) 2444 feet.

(11.) $\dfrac{75}{176\pi}\sqrt{\dfrac{805}{39}}$ miles an hour. (12.) $\dfrac{172,800\pi^2}{a.n(2880-a)}$.

MISCELLANEOUS.

(1.) $1:2$. (2.) $\dfrac{125}{gl}$ ounces.

(5.) The highest must start $\sqrt{\dfrac{8}{g}}(\sqrt{h}-\sqrt{h'})$ seconds before the lowest, h and h' being their initial heights above the plane.

(10). Inclination to horizon is $\sin^{-1}\dfrac{4}{5}$. Horizontal component is $\dfrac{3}{5}$ of whole. (13.) $\dfrac{1}{5}$, $\dfrac{1}{150}$, $\dfrac{25}{36}$. (14.) $\dfrac{1}{6}$ seconds nearly.

(25.) $\dfrac{av\pm\sqrt{c^2(v^2+v'^2)-a^2v'^2}}{v^2+v'^2}$, when v, v' are the velocities.

(35.) $m'+m'' : 2m+2m'$, m, m', m'' being respectively the masses of frog, frog's bucket, other bucket.

(36.) $\dfrac{2\pi}{t^2}\sqrt{\dfrac{r}{g}}$. (38.) $\dfrac{1}{800}$ seconds.

THE END.

Edinburgh University Press:
T AND A. CONSTABLE, PRINTERS TO HER MAJESTY.

3, WATERLOO PLACE, PALL MALL.
January, 1876.

𝕭ooks for 𝕾chools and 𝕮olleges

PUBLISHED BY

MESSRS. RIVINGTON

HISTORY

An English History for the Use of Public Schools.

By *the* Rev. J. FRANCK BRIGHT, M.A., *Fellow of University College,
and Historical Lecturer in Balliol, New, and University Colleges, Oxford;*
late Master of the Modern *School at Marlborough College.*

With numerous Maps and Plans. Crown 8vo.

This work is divided into three Periods of convenient and handy size, especially adapted for use in Schools, as well as for Students reading special portions of History for local and other Examinations. It will also be issued in one complete Volume.

Period I.—MEDIÆVAL MONARCHY: The departure of the Romans, to Richard III. From A.D. 449 to A.D. 1485. 4*s.* 6*d.*

Period II.—PERSONAL MONARCHY: Henry VII. to James II. From A.D. 1485 to A.D. 1688. [*Just Ready.*

Period III.—CONSTITUTIONAL MONARCHY: William and Mary, to the present time. From A.D. 1688 to A.D. 1837. [*Nearly Ready.*

About five years ago, **after a** meeting of a considerable number of Public School Masters, it was proposed to the Author that he should write a School History of England. As the suggestion was generally supported he undertook the task. The work was intended **to** supply some deficiencies felt to exist in the School Books which were at that time procurable. It was hoped that the work would be completed in three years, but a series of untoward events has postponed its completion till now. The Author has attempted to embody, in the present publication, so much of the fruit of many years' historical reading, and of considerable experience in teaching history, as he believes will be useful in rendering the study at once **an** instructive and an interesting pursuit for boys. Starting from **the supposition that** his readers know but little of the subject, he has tried to give a plain **narrative of events,** and at the same time so far to trace their connection, causes, and effects, as to supply **the** student with a more reasonable and intelligent idea of the course of English History than is given by any mere compendium of facts. It has been thought convenient to retain the ordinary divisions into reigns, and to follow primarily, throughout, the Political History of the country; at the same time considerable **care** has been given **to bring out the great** Social Changes which have occurred from **time** to time, and **to** follow the growth of the people and nation at large, as well as that of the Monarchy or of special classes. A considerable number of genealogies of the leading Houses of the 14th and 15th centuries have been introduced to illustrate that period. The later periods are related at considerably greater length than the earlier ones. The foreign events in which England took part have been, as far as space allowed, brought into due prominence; while by the addition of numerous maps and plans, in which every name mentioned will be found, it is hoped that reference to a separate atlas will be found unnecessary. The marginal analysis has been collected at the beginning of the volume, so as to form an abstract of the History, suitable for the **use of those who are** beginning the study.

(*See Specimen Page, No.* 1.)

LONDON, OXFORD, AND CAMBRIDGE.

HISTORICAL HANDBOOKS

Edited by

OSCAR BROWNING, M.A.,

FELLOW OF KING'S COLLEGE, CAMBRIDGE; ASSISTANT-MASTER **AT ETON COLLEGE.**

Small 8vo.

(See Specimen Pages, Nos. 3 and 4.)

HISTORY OF THE ENGLISH INSTITUTIONS.

By PHILIP V. SMITH, M.A., *Barrister-at-Law; Fellow of King's College, Cambridge.* 3s. 6d.

CONTENTS.

Social and Local Development of the Constitution.

Origin of the English Institutions—The People—Local Government.

Constituents of the Central Authority.

The King—Parliament—The King's Council.

Central Government.

Legislation—Judicature—The Executive—Taxation—Chronological Table.
Index and Glossary.

"It contains in a short compass an amount of information not otherwise accessible to students without considerable research. The chapter on Local Government in particular is well executed. It would be hard to name any other book in which the history of our local institutions, from the Gemots of the first Teutonic settlers down to the County Court, the Local Government Board, and the School Board of our own day, is to be found."

Athenæum.

HISTORY OF FRENCH LITERATURE.

Adapted from the French of M. DEMOGEOT, *by* C. BRIDGE.
3s. 6d.

"An excellent manual."—*Athenæum.*
"A clever adaptation."—*London Quarterly Review.*
"It is clear, idiomatic, and flowing, possessing all the characteristics of good English composition. Its perusal will furnish abundant evidence of the richness and variety of French literature, of which it is a good and sufficient handbook."—*British Quarterly Review.*
"We cannot too highly commend this careful analysis of the characteristics of the great French writer."—*Standard.*
"Unlike most manuals, it is readable as well as accurate."—*Echo.*

HISTORICAL HANDBOOKS—continued.

THE ROMAN EMPIRE. A.D. 395–800.

By A. M. CURTEIS, M.A., *Assistant-Master at Sherborne School,
late Fellow of Trinity College, Oxford.*

With Maps. 3s. 6d.

CONTENTS.

Administrative and Legal Unity—The Christian Church in the **First Four** Centuries—
The Barbarians on **the** Frontier—Century IV.—Church and State in Constantinople,
Eutropius and Chrysostom—Chrysostom and the Empress Eudoxia—Alaric and the Visi-
goths, 396–419—Genseric and the Vandals, 423–533—Attila and the Huns, 435–453—
The "Change of Government," commonly called the Fall of the Western Empire, 475–526
—The Emperor Justinian, 527–565—The Empire in relation to the Barbarians of the East,
450–650—Mohammed and Mohammedanism, 622–711—The Popes and the Lombards in
Italy, 540–740—The Franks and the Papacy, 500–800—Synopsis of Historical Events—
Index.

MAPS.

Central Europe, **about** A.D. **400.**—The **Roman** Empire **at** the beginning **of the** sixth
Century.—Italy, 600–750.—Europe in the time of Charles the Great.

"**We** have very carefully examined the
chapters on the ' Barbarians,' the Visigoths,
the Vandals, and the Huns, and can pro-
nounce them the best condensed account
that we have read of the westerly migra-
tions."—*Athenæum.*

" An admirable specimen of careful con-
densation and good arrangement, and as a
school book it will assuredly possess a high
value."—*Scotsman.*

"The period with which it deals is
neglected in schools for want of text-books,
but is full of most important historical

teaching. Mr. Curteis' little book is admirably written for teaching purposes; it is
clear, definite, well-arranged, and interesting."—*Academy.*

"Appears to be a good school book for
the higher forms."—*Westminster Review.*

" Will prove of great service to students,
and we commend it to the notice of those
who intend competing in the Civil Service
Examinations. Mr. Curteis has executed
his task with great care and judgment."
Civil Service Gazette.

HISTORY OF MODERN ENGLISH LAW.

By Sir ROLAND KNYVET WILSON, Bart., M.A., *Barrister-at-Law ;
late Fellow of King's College, Cambridge.*

3s. 6d.

CONTENTS.

The English Law in the Time of Blackstone.

The Form in which it was Enunciated—Property, Contract, and Absolute Duties—
Wrongs and Remedies, Civil and Criminal—Procedure—Laws relating to Special
Classes of Persons.

Life and Work of Bentham.

Biography—The Writings of Bentham—Early Attempts at Law Reform.

Legal Changes since 1825.

Changes as to the Form in which the Law is Enunciated—Changes in the Law of
Property, Contract, and Absolute Duties—Changes in the Laws as to Wrongs and
Remedies—Changes in the Laws relating to Procedure and Evidence—Changes
in the Laws relating to Special Classes of Persons—Chronological Table of Cases
and Statutes—Index.

[*HISTORICAL HANDBOOKS—Continued.*

LONDON, OXFORD, AND CAMBRIDGE.

HISTORICAL HANDBOOKS—continued.

English History in the XIVth Century.

By CHARLES H. PEARSON, M.A., *Principal of the Presbyterian Ladies' College, Melbourne, late Fellow of Oriel College, Oxford, and Professor of History in the University of Melbourne.*

[*In the Press.*

The Reign of Louis XI.

By F. WILLERT, M.A., *Fellow of Exeter College, Oxford.*

[*In the Press.*

The Great Rebellion.

By the EDITOR.

History of the French Revolution.

By the Rev. J. FRANCK BRIGHT, M.A., *Fellow of University College, and Historical Lecturer in Balliol, New, and University Colleges, Oxford; late Master of the Modern School at Marlborough College.*

The Age of Chatham.

By Sir W. R. ANSON, Bart., M.A., *Fellow of All Souls' College, and Vinerian Reader of Law, Oxford.*

The Age of Pitt.

By the Same.

The Supremacy of Athens.

By R. C. JEBB, M.A., *Fellow and Tutor of Trinity College, Cambridge, and Public Orator of the University.*

The Roman Revolution. From B.C. 133 to the Battle of Actium.

By H. F. PELHAM, M.A., *Fellow and Lecturer of Exeter College, Oxford.*

History of the United States.

By SIR GEORGE YOUNG, BART., M.A., *late Fellow of Trinity College, Cambridge.*

History of Roman Political Institutions.

By J. S. REID, M.L., *Christ's College, Cambridge.*

LONDON, OXFORD, AND CAMBRIDGE.

In the Press

HISTORICAL BIOGRAPHIES

Edited by

THE REV. M. CREIGHTON, M.A.,

FELLOW AND TUTOR OF MERTON COLLEGE, OXFORD,

With Frontispiece and Maps.

The most important and the most difficult point in Historical Teaching is to awaken a real interest in the minds of Beginners. For this purpose concise handbooks are seldom useful. General sketches, however accurate in their outlines of political or constitutional development, and however well adapted to dispel false ideas, still do not make history a living thing to the *young*. They are most valuable as maps on which to trace **the** route beforehand and show its direction, but they will seldom allure any one to take a walk.

The object of this series of Historical Biographies is to try and select from English History a few men whose lives were lived in stirring times. The intention is to treat their lives and times in some little detail, and to group round them the most distinctive features of the periods before and after those in which they lived.

It is hoped that in this way interest **may be a**wakened without any sacrifice of accuracy, and that personal sympathies may be kindled without forgetfulness of the principles involved.

It may be added **that** round the lives of individuals it will be possible to bring together facts of social life in a clearer way, and to reproduce a more vivid picture of particular times than is possible in a historical handbook.

By reading Short Biographies a few clear ideas may be formed in the pupil's mind, which may stimulate to further reading. A vivid impression of one period, however short, will carry the pupil onward and give more general histories an interest in their turn. Something, at least, will be gained if the pupil realises that men in past times lived and moved in the same sort of way as they do at present.

It is proposed to issue the following Biographies **adapted** to the reading of pupils between the ages of **12** and **15**:

1. SIMON DE MONTFORT.	5. THE DUKE OF MARLBOROUGH.
2. THE BLACK PRINCE.	6. WILLIAM PITT,
3. SIR WALTER RALEIGH.	or, THE DUKE OF WELLINGTON.
4. OLIVER CROMWELL.	

History of the Church under the Roman Empire, A.D. 30-476.

By the Rev. A. D. CRAKE, B.A., *Chaplain of All Saints' School, Bloxham.*

Crown 8vo. 7s. 6d.

A History of England for Children.

By GEORGE DAVYS, D.D., *formerly Bishop of Peterborough.*

New Edition. 18mo. 1s. 6d.

With twelve Coloured Illustrations. Square cr. 8vo. 3s. 6d.

LONDON, OXFORD, AND CAMBRIDGE.

ENGLISH

ENGLISH SCHOOL-CLASSICS

With Introductions, and Notes at the end of each Book.

Edited by FRANCIS STORR, B.A.,

CHIEF MASTER OF MODERN SUBJECTS IN MERCHANT TAYLORS' SCHOOL, LATE SCHOLAR
OF TRINITY COLLEGE, CAMBRIDGE, AND BELL UNIVERSITY SCHOLAR.

Small 8vo.

THOMSON'S SEASONS: Winter.
With Introduction to the Series, by the Rev. J. FRANCK BRIGHT, M.A., Fellow of University College, and Historical Lecturer in Balliol, New, and University Colleges, Oxford; late Master of the Modern School at Marlborough College. 1s.

COWPER'S TASK.
By FRANCIS STORR, B.A., Chief Master of Modern Subjects in Merchant Taylors' School. 2s.
Part I. (Book I.—The Sofa; Book II.—The Timepiece) 9d. Part II. (Book III. —The Garden; Book IV.—The Winter Evening) 9d. Part III. (Book V.—The Winter Morning Walk; Book VI.—The Winter Walk at Noon) 9d.

SCOTT'S LAY OF THE LAST MINSTREL.
By J. SURTEES PHILLPOTTS, M.A., Head Master of Bedford School, formerly Fellow of New College, Oxford. 2s. 6d.
Part I. (Canto I., with Introduction, &c.) 9d. Part II. (Cantos II. and III.) 9d. Part III. (Cantos IV. and V.) 9d. Part IV. (Canto VI.) 9d.

SCOTT'S LADY OF THE LAKE.
By R. W. TAYLOR, M.A., Assistant-Master at Rugby School. 2s.
Part I. (Cantos I. and II.) 9d. Part II. (Cantos III. and IV.) 9d. Part III. (Cantos V. and VI.) 9d.

NOTES TO SCOTT'S WAVERLEY.
By H. W. EVE, M.A., Assistant-Master at Wellington College. 1s., or with the Text, 2s. 6d.

TWENTY OF BACON'S ESSAYS.
By FRANCIS STORR, B.A., Chief Master of Modern Subjects in Merchant Taylors' School. 1s.

SIMPLE POEMS.
Edited by W. E. MULLINS, M.A. Assistant-Master at Marlborough College. 8d.

SELECTIONS FROM WORDSWORTH'S POEMS.
By H. H. TURNER, B.A., late Scholar of Trinity College, Cambridge. 1s.

WORDSWORTH'S EXCURSION: The Wanderer.
By H. H. TURNER, B.A., late Scholar of Trinity College, Cambridge. 1s.

MILTON'S PARADISE LOST.
By FRANCIS STORR, B.A., Chief Master of Modern Subjects in Merchant Taylors' School.
Book I. 9d. Book II. 9d.

ENGLISH SCHOOL CLASSICS—continued.

SELECTIONS FROM THE SPECTATOR.
By OSMUND AIRY, M.A., Assistant-Master at Wellington College. 1*s.*

BROWNE'S RELIGIO MEDICI.
By W. P. SMITH, M.A., Assistant-Master at Winchester College. 1*s.*

GOLDSMITH'S TRAVELLER AND DESERTED VILLAGE.
By C. SANKEY, M.A., Assistant-Master at Marlborough College. 1*s.*

EXTRACTS FROM GOLDSMITH'S VICAR OF WAKEFIELD.
By C. SANKEY, M.A., Assistant-Master at Marlborough College. 1*s.*

SELECTIONS FROM BURNS' POEMS.
By A. M. BELL, M.A., Balliol College, Oxford.

MACAULAY'S ESSAYS.
MOORE'S LIFE OF BYRON. By FRANCIS STORR, B.A. 9*d.*
BOSWELL'S LIFE OF JOHNSON. By FRANCIS STORR, B.A. 9*d.*
HALLAM'S CONSTITUTIONAL HISTORY. By H. F. BOYD, late Scholar of
Brasenose College, Oxford. 1*s.*

SOUTHEY'S LIFE OF NELSON.
By W. E. MULLINS, M.A., Assistant-Master at Marlborough College.

** *The General Introduction to the Series will be found in* Thomson's WINTER.

(*See Specimen Pages, Nos.* 5 *and* 6.)

OPINIONS OF TUTORS AND SCHOOLMASTERS.

"Nothing can be better than the idea and the execution of the English School-Classics, edited by Mr. Storr. Their cheapness and excellence encourage us to the hope that the study of our own language, too long neglected in our schools, may take its proper place in our curriculum, and may be the means of inspiring that taste for literature which it is one of the chief objects of education to give, and which is apt to be lost sight of in the modern style of teaching Greek and Latin Classics with a view to success in examinations."—*Oscar Browning, M.A., Fellow of King's College, Cambridge; Assistant-Master at Eton College.*

"I think the plan of them is excellent; and those volumes which I have used I have found carefully and judiciously edited, neither passing over difficulties, nor preventing thought and work on the pupil's part by excessive annotation."—*Rev. C. B. Hutchinson, M.A., Assistant-Master in Rugby School.*

"I think that these books are likely to prove most valuable. There is great variety in the choice of authors. The notes seem sensible, as far as I have been able to examine them, and give just enough help, and not too much; and the size of each volume is so small, that in most cases it need not form more than one term's work.

Something of the kind was greatly wanted."—*E. E. Bowen, M.A., Master of the Modern Side, Harrow School.*

"I have used some of the volumes of your English School-Classics for several months in my ordinary form work, and I have recommended others to be set as subjects for different examinations for which the boys have to prepare themselves. I shall certainly continue to use them, as I have found them to be very well suited to the wants of my form."—*C. M. Bull, M.A., Master of the Modern School in Marlborough College.*

"I have no hesitation in saying that the volumes of your Series which I have examined appear to me far better adapted for school use than any others which have come under my notice. The notes are sufficiently full to supply all the information which a boy needs to understand the text without superseding the necessity of his thinking. The occasional questions call the learner's attention to points which he can decide from his own resources. The general plan, and the execution of the volumes which have come before me, leave little to be desired in a School Edition of the English Classics."—*The Rev. Chas. Grant Chittenden, M.A., The Grange, Hoddesdon, Herts.*

LONDON, OXFORD, AND CAMBRIDGE.

SELECT PLAYS OF SHAKSPERE

RUGBY EDITION.

With an Introduction and Notes to each Play.

Small 8vo.

As You Like It. 2s.   Hamlet. 2s. 6d.

Macbeth. 2s.   King Lear.

Edited by the Rev. Charles E. Moberly, M.A., *Assistant-Master at* Rugby *School, and formerly Scholar of Balliol College, Oxford.*

Coriolanus. 2s. 6d.

Edited by Robert Whitelaw, M.A., *Assistant-Master at Rugby School, formerly Fellow of Trinity College, Cambridge.*

The Tempest.

Edited by J. Surtees Phillpotts, M.A., *Head-Master of Bedford Grammar School, formerly Fellow of New College, Oxford.*

With Notes at the end of the Volume.   [*Nearly Ready.*

The Merchant of Venice.

Edited by R. W. Taylor, M.A., *Assistant-Master at Rugby School.*

With Notes at the end of the Volume.

[*In preparation.*

Reflections on the Revolution in France in 1790.

By the Right Hon. Edmund Burke, M.P.

New Edition, with a short Biographical Notice. Crown 8vo. 3s. 6d.

A Practical Introduction to English Prose Composition.

An English Grammar for Classical Schools, with Questions, and a Course of Exercises.

By Thomas Kerchever Arnold, M.A.

Tenth Edition. 12mo. 4s. 6d.

MATHEMATICS

RIVINGTONS' MATHEMATICAL SERIES

The following Schools, amongst many others, use this Series :— Eton : Harrow : Winchester : Charterhouse : Marlborough : Shrewsbury : Cheltenham : Clifton : City of London School : Haileybury : Tonbridge : Durham : Fettes College, Edinburgh : Owen's College, Manchester : H.M.'s Dockyard School, Sheerness : Hurstpierpoint : King William's College, Isle of Man : St. Peter's, Clifton, York : Birmingham : Bedford : Felsted : **Christ's** College, **Finchley :** Liverpool College : Windermere College : Eastbourne College : Brentwood : Perse School, Cambridge. Also in **use** in Canada : Harvard College, U.S. : H.M. Training Ships : Royal Naval College, Greenwich : Melbourne University, Australia : the other Colonies : **and** some of the Government **Schools in India.**

OPINIONS OF TUTORS AND SCHOOLMASTERS.

" A person who carefully studies these books will have a thorough and accurate knowledge of the subjects on which they treat."—*H. A. Morgan, M.A., Tutor of Jesus College, Cambridge.*

"We have for some time used your Mathematical books in our Lecture Room, and find them well arranged, and well calculated to clear up the difficulties of the subjects. The examples also are numerous and well-selected."—*N. M. Ferrers, M.A., Fellow and Tutor of Gonville and Caius College, Cambridge.*

" I have used in my Lecture Room Mr. Hamblin Smith's text-books with very great advantage."—*James Porter, M.A., Fellow and Tutor of St. Peter's College, Cambridge.*

" For beginners there could be no better books, as I have found when examining different schools."—*A.W.W. Steel, M.A., Fellow and Assistant-Tutor of Gonville and Caius College, Cambridge.*

" I consider Mr. Hamblin **Smith's** Mathematical **Works** to be a very **valuable** series for **beginners.** His Algebra **in** particular I think is the best book of its **kind** for schools and for the ordinary course at Cambridge."

F. Patrick, M.A., Fellow and Tutor of Magdalen College, Cambridge.

"The series is a model of clearness and insight into possible difficulties."—*Rev. J. F. Blake, St. Peter's College, Clifton, York.*

" I can say with pleasure that I have used your books extensively in my work at Haileybury, **and have** found them on the whole well **adapted for** boys."—*Thomas Pitts, M.A., Assistant Mathematical Master at Haileybury College.*

" I can strongly recommend them all."
W. Henry, M.A., Sub-Warden, Trinity College, Glenalmond.

" I consider Mr. Smith has supplied a great want, and cannot but think that his works must command extensive use in good schools."—*J. Henry, B.A., Head-Master, H.M. Dockyard School, Sheerness, and Instructor of Engineers, R.N.*

"We have used your Algebra and Trigonometry extensively at this School from the time they were first published, and I thoroughly agree with every mathematical teacher I have met, that, as school text-books, they **have no** equals. **We** are introducing your Euclid gradually into the School."—*Rev. D. Edwards, sen., Mathematical Master at the College, Hurstpierpoint, Sussex.*

" I consider **them to** be the best books of their kind on the subject which I have yet seen."—*Joshua Jones, D.C.L., Head-Master, King William's College, Isle of Man.*

" I have very great pleasure in expressing an opinion as to the value of these books. I have used them under very different circumstances, and have always been satisfied with the results obtained." — *C. H. W. Biggs, Mathematical Editor of the 'English Mechanic,' Editor of the 'Monthly Journal of Education.'*

ELEMENTARY ALGEBRA.

By J. HAMBLIN SMITH, M.A., *of Gonville and Caius College, and late Lecturer at St. Peter's College, Cambridge.*

12mo. 3s. Without Answers, 2s. 6d.

A KEY TO ELEMENTARY ALGEBRA.

Crown 8vo. 9s.

EXERCISES ON ALGEBRA.

By J. HAMBLIN SMITH, M.A.

12mo. 2s. 6d.

Copies may be had without the Answers.

ALGEBRA.   Part II.

By E. J. GROSS, M.A., *Fellow of Gonville and Caius College, Cambridge.*

Crown 8vo. 8s. 6d.

"We have to congratulate Mr. Gross on his excellent treatment of the more difficult chapters in Elementary Algebra. His work satisfies not only in every respect the requirements of a first-rate text-book on the subject, but is not open to the standing reproach of most English mathematical treatises for students, a minimum of teaching and a maximum of problems. The hard work and considerable thought which Mr. Gross has devoted to the book will be seen on every page by the experienced teacher; there is not a word too much, nor is the student left without genuine assistance where it is needful. The language is precise, clear, and to the point. The problems are not too numerous, and selected with much tact and judgment. The range of the book has been very rightly somewhat extended beyond that assigned to simpler treatises, and it includes the elementary principles of Determinants. This chapter especially will be read with satisfaction by earnest students, and the mode of exposition will certainly have the approval of teachers. Altogether we think that this *Algebra* will soon become a general text-book, and will remain so for a long time to come." — *Westminster Review.*

KINEMATICS AND KINETICS.

By E. J. GROSS, M.A.

Crown 8vo.   [*Just ready.*

A TREATISE ON ARITHMETIC.

By J. HAMBLIN SMITH, M.A.

12mo. 3s. 6d.

(*See Specimen Page, No. 7.*)

A KEY TO ARITHMETIC.

Crown 8vo.   [*In the Press.*

ELEMENTS OF GEOMETRY.

By J. HAMBLIN SMITH, M.A.

12mo. 3s. 6d.

Containing Books 1 to 6, and portions of Books 11 and 12, of EUCLID, with Exercises and Notes, arranged with the Abbreviations admitted in the Cambridge Examinations.

Part I., containing Books 1 and 2 of Euclid, limp cloth, 1s. 6d., may be had separately.

(See Specimen Page, No. 8.)

GEOMETRICAL CONIC SECTIONS.

By G. RICHARDSON, M.A., *Assistant-Master at Winchester College, and late Fellow of St. John's College, Cambridge.*

Crown 8vo. 4s. 6d.

TRIGONOMETRY.

By J. HAMBLIN SMITH, M.A.

12mo. 4s. 6d.

ELEMENTARY STATICS.

By J. HAMBLIN SMITH, M.A.

12mo. 3s.

ELEMENTARY HYDROSTATICS.

By J. HAMBLIN SMITH, M.A.

12mo. 3s.

BOOK OF ENUNCIATIONS

FOR HAMBLIN SMITH'S GEOMETRY, ALGEBRA, TRIGONOMETRY, STATICS, AND HYDROSTATICS.

12mo. 1s.

Arithmetic, Theoretical and Practical.

By W. H. GIRDLESTONE, M.A., *of Christ's College, Cambridge, Principal of the Theological College, Gloucester.*

New Edition. Crown 8vo. 6s. 6d.
Also a School Edition. Small 8vo. 3s. 6d.

LONDON, OXFORD, AND CAMBRIDGE.

SCIENCE

Preparing for Publication,

SCIENCE CLASS-BOOKS

Edited by

The Rev. ARTHUR RIGG, M.A.,

LATE PRINCIPAL OF THE COLLEGE, CHESTER.

These Volumes are designed expressly for School use, and by their especial reference to the requirements of a School Class-Book, aim at making Science-teaching a subject for regular and methodical study in Public and Private Schools.

AN ELEMENTARY CLASS-BOOK ON SOUND.

By GEORGE CAREY FOSTER, B.A., F.R.S., *Fellow of, and Professor of Physics in,* University College, London.

AN ELEMENTARY CLASS-BOOK ON ELECTRICITY.

By GEORGE CAREY FOSTER, B.A., F.R.S., *Fellow of, and Professor of Physics in, University College, London.*

BOTANY FOR CLASS-TEACHING.

With Exercises for Private Work.

By F. E. KITCHENER, M.A., F.L.S., *Assistant-Master at Rugby School, and late Fellow of Trinity College, Cambridge.*

ASTRONOMY FOR CLASS-TEACHING.

With Exercises for Private Work.

By WALLIS HAY LAVERTY, M.A., *late Fellow of Queen's College, Oxford.*

The knowledge of Mathematics assumed will be Euclid, Books I.-VI., and Quadratic Equations.

Other Works are in preparation.

LONDON, OXFORD, AND CAMBRIDGE.

A Year's Botany.

Adapted to Home and School Use.
By FRANCES ANNA KITCHENER.
Illustrated by the Author. Crown 8vo. 5*s.*
(*See Specimen Page, No.* 2.)

CONTENTS.

General Description of Flowers—Flowers with Simple Pistils—Flowers with Compound Pistils—Flowers with Apocarpous Fruits—Flowers with Syncarpous Fruits—Stamens and Morphology of Branches—Fertilisation—Seeds—Early Growth and Food of Plants—Wood, Stems, and Roots—Leaves—Classification—Umbellates, Composites, Spurges, and Pines—Some Monocotyledonous Families—Orchids—Appendix of Technical Terms—Index.

An Easy Introduction to Chemistry.

For the use of Schools.
Edited by the Rev. ARTHUR RIGG, M.A., *late Principal of The College, Chester, and* WALTER T. GOOLDEN, B.A., *late Science Scholar of Merton College, Oxford.*
New Edition, revised. With Illustrations. Crown 8vo. 2*s.* 6*d.*
(*See Specimen Page, No.* 9.)

"We seldom come across a work of such simplicity in chemistry as this. It ought to be in the hands of every student of chemistry."—*Chemical Review.*

"There are a simplicity and a clearness in the description and explanations given in this little volume which certainly commend it to the attention of the young." *Athenæum.*

Notes on Building Construction.

Arranged to meet the requirements of the syllabus of the Science and Art Department of the Committee of Council on Education, South Kensington Museum.
PART I.—FIRST STAGE, OR ELEMENTARY COURSE.
PARTS II. and III. [*In the Press.*
Medium 8vo, with 325 woodcuts, 10*s.* 6*d.*

REPORT ON THE EXAMINATION IN BUILDING CONSTRUCTION, HELD BY THE SCIENCE AND ART DEPARTMENT, SOUTH KENSINGTON, IN MAY, 1875.—"The want of a text-book in this subject, arranged in accordance with the published syllabus, and therefore limiting the students and teachers to the prescribed course, has lately been well met by a work published by Messrs. Rivingtons, entitled '*Notes on Building Construction,* arranged to meet the requirements of the Syllabus of the Science and Art Department of the Committee of Council on Education, South Kensington.'
June 18, 1875. (Signed) H. C. SEDDON, Major, R.E."

"Something of the sort was very much needed. A book distilling the substance of larger works and putting the outlines of constructional science together in a small compass, is a very important aid to students. A very useful little book."—*Builder.*

"The text is prepared in an extremely simple and consecutive manner, advancing from rudimental and general statements to those which are comparatively advanced : it is a thoroughly coherent, self-sustained account."—*Athenæum.*

LONDON, OXFORD, AND CAMBRIDGE.

LATIN

Easy Latin Stories for Beginners.

Forming a First Latin Reading Book for Junior Forms in Schools.
With Notes and a Vocabulary.

By G. L. BENNETT, M.A., *Assistant Master at Rugby School.*

Crown 8vo. [*In the press.*

Elementary Rules of Latin Pronunciation.

By ARTHUR HOLMES, M.A., *late Senior Fellow and Dean of Clare College, Cambridge.*

Crown 8vo. On a card, 9*d.*

Outlines of Latin Sentence Construction.

By E. D. MANSFIELD, B.A., *Assistant-Master at Clifton College.*

Demy 8vo. On a card, 1*s.*

Easy Exercises in Latin Prose.

By CHARLES BIGG, M.A., *Principal of Brighton College.*

Small 8vo. 1*s.* 4*d.* ; sewed, 9*d.*

Latin Prose Exercises.

For Beginners, and Junior Forms of Schools.

By R. PROWDE SMITH, B.A., *Assist.-Master at Cheltenham College.*

New Edition. Crown 8vo. 2*s.* 6*d.*

An Elementary Latin Grammar.

By J. HAMBLIN SMITH, M.A., *of Gonville and Caius College, Cambridge ; late Lecturer of S. Peter's College, Cambridge.*

Small 8vo. 3*s.* 6*d.*

LONDON, OXFORD, AND CAMBRIDGE.

Henry's First Latin Book.

By THOMAS KERCHEVER ARNOLD, M.A.

Twenty-second Edition.   12mo.   3s.   Tutor's Key, 1s.

Recommended in the *Guide to the Choice of Classical Books* by J. B. Mayor, M.A., Professor of Classical Literature at King's College, late Fellow and Tutor of St. John's College, Cambridge.

A Practical Introduction to Latin Prose Composition.

By THOMAS KERCHEVER ARNOLD, M.A.

Sixteenth Edition.   8vo.   6s. 6d.   Tutor's Key, 1s. 6d.

Cornelius Nepos.

With Critical Questions and Answers, and an Imitative Exercise on each Chapter.

By THOMAS KERCHEVER ARNOLD, M.A.

Fifth Edition.   12mo.   4s.

A First Verse Book.

Being an Easy Introduction to the Mechanism of the Latin Hexameter and Pentameter.

By THOMAS KERCHEVER ARNOLD, M.A.

Eleventh Edition.   12mo.   2s.   Tutor's Key, 1s.

Progressive Exercises in Latin Elegiac Verse.

By C. G. GEPP, B.A., *late Junior Student of Christ Church, Oxford; Head-Master of the College, Stratford-on-Avon.*

Third Edition, Revised.   Crown 8vo.   3s. 6d.   Tutor's Key, 5s.

Recommended in the *Guide to the Choice of Classical Books* by J. B. Mayor, M.A., Professor of Classical Literature at King's College, late Fellow and Tutor of St. John's College, Cambridge.

Selections from Livy, Books VIII. and IX.

With Notes and Map.

By E. CALVERT, LL.D., *St. John's College, Cambridge; and* R. SAWARD, M.A., *Fellow of St. John's College, Cambridge; Assistant-Master in Shrewsbury School.*

Small 8vo.   2s.

LONDON, OXFORD, AND CAMBRIDGE.

New Edition, re-arranged, with fresh Pieces and additional References.

Materials and Models for Latin Prose Composition.

Selected and arranged by J. Y. SARGENT, M.A., *Fellow and Tutor of Magdalen College, Oxford;* and T. F. DALLIN, M.A., *Tutor, late Fellow, of Queen's College, Oxford.*

Crown 8vo. 6s. 6d.

(*See Specimen Page, No.* 10.)

Latin Version of (60) Selected Pieces from Materials and Models.

By J. Y. SARGENT, M.A.

Crown 8vo. 5s.

May be had by Tutors only, on direct application to the Publishers.

Stories from Ovid in Elegiac Verse.

With Notes for School Use and Marginal References to the PUBLIC SCHOOL LATIN PRIMER.

By R. W. TAYLOR, M.A., *Assistant-Master at Rugby School, late Fellow of St. John's College, Cambridge.*

Crown 8vo. 3s. 6d.

(*See Specimen Pages, Nos.* 11 *and* 12.)

The Æneid of Virgil.

Edited, with Notes at the end, by FRANCIS STORR, B.A., *Chief Master of Modern Subjects in Merchant Taylors' School.*

BOOKS XI and XII.

Crown 8vo. 2s. 6d.

(*See Specimen Pages, Nos.* 13 *and* 14.)

Classical Examination Papers.

Edited, with Notes and References, by P. J. F. GANTILLON, M.A., *Classical Master in Cheltenham College.*

Crown 8vo. 7s. 6d.

Or interleaved with writing-paper, half-bound, 10s. 6d.

Eclogæ Ovidianæ.

From the Elegiac Poems. With English Notes.

By THOMAS KERCHEVER ARNOLD, M.A.

Thirteenth Edition. 12mo. 2s. 6d.

Terenti Comoediae.

Edited by T. L. PAPILLON, M.A., *Fellow of New College, and late Fellow of Merton, Oxford.*
ANDRIA ET EUNUCHUS. 4*s.* 6*d.*
ANDRIA. New Edition, with Introduction on Prosody. 3*s.* 6*d.*
Crown 8vo.
Forming a Part of the " Catena Classicorum."

Juvenalis Satirae.

Edited by G. A. SIMCOX, M.A., *Fellow of Queen's College, Oxford.*
THIRTEEN SATIRES.
Second Edition, enlarged and revised. Crown 8vo. 5*s.*
Forming a Part of the " Catena Classicorum."

Persii Satirae.

Edited by A. PRETOR, M.A., *of Trinity College, Cambridge, Classical Lecturer of Trinity Hall, Composition Lecturer of the Perse Grammar School, Cambridge.*
Crown 8vo. 3*s.* 6*d.*
Forming a Part of the " Catena Classicorum."

Horati Opera.

By J. M. MARSHALL, M.A., *Under-Master in Dulwich College.*
VOL. I.—THE ODES, CARMEN SECULARE, **AND** EPODES.
Crown 8vo. 7*s.* 6*d.*
Forming a Part of the " Catena Classicorum."

Taciti Historiae. BOOKS I. and II.

Edited by W. H. SIMCOX, M.A., *Fellow of Queen's College, Oxford.*
Crown 8vo. 6*s.*
Forming a Part of the " Catena Classicorum."

Taciti Historiae. BOOKS III. IV. and V.

Edited by W. H. SIMCOX, M.A., *Fellow of Queen's College, Oxford.*
Crown 8vo. [*In the Press.*
Forming a Part of the "Catena Classicorum."

GREEK

An Elementary Grammar for the Use of Beginners.

By EVELYN ABBOTT, M.A., *Lecturer in Balliol College, Oxford, and late Assistant-Master in Clifton College.*   [*In the Press.*

Elements of Greek Accidence.

By EVELYN ABBOTT, M.A., *Lecturer in Balliol College, Oxford, and late Assistant-Master in Clifton College.*
Crown 8vo.   4s. 6d.

"This is an excellent book. The compilers of elementary Greek Grammars have not before, so far as we are aware, made full use of the results obtained by the labours of philologists during the last twenty-five years. Mr. Abbott's great merit is that he has; and a comparison between his book and the *Rudimenta* of the late Dr. Donaldson—a most excellent volume for the time at which it was published—will show how considerable the advance has been; while a comparison with the works in ordinary use, which have never attained anything like the standard reached by Dr. Donaldson, will really surprise the teacher."—*Athenæum.*

An Introduction to Greek Prose Composition.

By ARTHUR SIDGWICK, M.A., *Assistant-Master at Rugby School, and formerly Fellow of Trinity College, Cambridge.*

[*In the Press.*

Zeugma; or, Greek Steps from Primer to Author.

By the Rev. LANCELOT SANDERSON, M.A., *Principal of Elstree School, late Scholar of Clare College, Cambridge; and the* Rev. F. B. FIRMAN, M.A., *Assistant-Master at Elstree School, late Scholar of Jesus College, Cambridge.*
Small 8vo.   1s. 6d.

A Table of Irregular Greek Verbs.

Classified according to the arrangement of Curtius's Greek Grammar.
By FRANCIS STORR, B.A., *Chief-Master of Modern Subjects in Merchant Taylors' School, late Scholar of Trinity College, Cambridge, and Bell University Scholar.*
On a Card.   1s.

Selections from Lucian.

With English Notes.

By EVELYN ABBOTT, M.A., *Lecturer in Balliol College, Oxford, and late Assistant-Master in Clifton College.*

Small 8vo.   3s. 6d.

Alexander the Great in the Punjaub.

Adapted from Arrian, Book V.

An easy Greek Reading Book, with Notes at the end and a Map.

By the Rev. CHARLES E. MOBERLY, M.A., *Assistant-Master in Rugby School, and formerly Scholar of Balliol College, Oxford.*

Small 8vo.   2s.

Stories from Herodotus.

The Tales of Rhampsinitus and Polycrates, and the Battle of Marathon and the Alcmæonidae. *In Attic Greek.*

Adapted for use in Schools, by J. SURTEES PHILLPOTTS, M.A., *Head Master of Bedford School; formerly Fellow of New College, Oxford.*

Crown 8vo.   1s. 6d.

Iophon: an Introduction to the Art of Writing Greek Iambic Verses.

By the WRITER *of "Nuces" and "Lucretilis."*

Crown 8vo.   2s.

The First Greek Book.

On the plan of *Henry's First Latin Book.*

By THOMAS KERCHEVER ARNOLD, M.A.

Sixth Edition.   12mo.   5s.   Tutor's Key, 1s. 6d.

A Practical Introduction to Greek Accidence.

By THOMAS KERCHEVER ARNOLD, M.A.

Ninth Edition.   8vo.   5s. 6d.

A Practical Introduction to Greek Prose Composition.

By THOMAS KERCHEVER ARNOLD, M.A.

Twelfth Edition.   8vo.   5s. 6d.   Tutor's Key, 1s. 6d.

LONDON, OXFORD, AND CAMBRIDGE.

C 2

SCENES FROM GREEK PLAYS
RUGBY EDITION

Abridged and adapted for the use of Schools, by

ARTHUR SIDGWICK, M.A.,
ASSISTANT-MASTER AT RUGBY SCHOOL, AND FORMERLY FELLOW OF
TRINITY COLLEGE, CAMBRIDGE.

Small 8vo.   1s. 6d. each.

Aristophanes.

THE CLOUDS. THE FROGS. THE KNIGHTS. PLUTUS.

Euripides.

IPHIGENIA IN TAURIS. THE CYCLOPS. ION.
ELECTRA. ALCESTIS. BACCHÆ. HECUBA.

Recommended in the *Guide to the Choice of Classical Books,* by J. B. Mayor, M.A., Professor of Classical Literature at King's College, late Fellow and Tutor of St. John's College, Cambridge.

"Mr. Sidgwick has put on the title-pages of these modest little volumes the words 'Rugby Edition,' but we shall be much mistaken if they do not find a far wider circulation. The prefaces or introductions which Mr. Sidgwick has prefixed to his 'Scenes' tell the youthful student all that he need know about the play that he is taking in hand, and the parts chosen are those which give the general scope and drift of the action of the play."—*School Board Chronicle.*

"Each play is printed separately, on good paper, and in a neat and handy form. The difficult passages are explained by the notes appended, which are of a particularly useful and intelligible kind. In all respects this edition presents a very pleasing contrast to the German editions hitherto in general use, with their Latin explanatory notes—themselves often requiring explanation. A new feature in this edition, which deserves mention, is the insertion in English of the stage directions. By means of them and the argument prefixed, the study of the play is much simplified."—*Scotsman.*

"A short preface explains the action of the play in each case, and there are a few notes at the end which will clear up most of the difficulties likely to be met with by the young student."—*Educational Times.*

"Just the book to be put into the hands of boys who are reading Greek plays. They are carefully and judiciously edited, and form the most valuable aid to the study of the elements of Greek that we have seen for many a day. The Grammatical Indices are especially to be commended."—*Athenæum.*

"These editions afford exactly the kind of help that school-boys require, and are really excellent class-books. The notes, though very brief, are of much use and always to the point, and the arguments and arrangement of the text are equally good in their way."—*Standard.*

"Not professing to give whole dramas, with their customary admixture of iambics, trochaics, and choral odes, as pabulum for learners who can barely digest the level speeches and dialogues commonly confined to the first-named metre, he has arranged extracted scenes with much tact and skill, and set them before the pupil with all needful information in the shape of notes at the end of the book; besides which he has added a somewhat novel, but highly commendable and valuable feature—namely, appropriate headings to the commencement of each scene, and appropriate stage directions during its progress."—*Saturday Review.*

"These are attractive little books, novel in design and admirable in execution. . . . . It would hardly be possible to find a better introduction to Aristophanes for a young student than these little books afford."
London Quarterly Review.

Homer's Iliad.

Edited, with Notes at the end, by J. SURTEES PHILLPOTTS, M.A., *Head Master of Bedford Grammar School, formerly Fellow of New College, Oxford.*

BOOK VI. Crown 8vo.

Homer for Beginners.

ILIAD, Books I.—III. With English Notes.
By THOMAS KERCHEVER ARNOLD, M.A.

Fourth Edition. 12mo. 3s. 6d.

The Iliad of Homer.

From the Text of Dindorf. With Preface and Notes.
By S. H. REYNOLDS, M.A., *Fellow and Tutor of Brasenose College, Oxford.*

Books I.—XII. Crown 8vo. 6s.
Forming a Part of the "Catena Classicorum."

The Iliad of Homer.

With English Notes and Grammatical References.
By THOMAS KERCHEVER ARNOLD, M.A.

Fifth Edition. 12mo. Half-bound, 12s.

A Complete Greek and English Lexicon for the Poems of Homer and the Homeridæ.

By G. CH. CRUSIUS. *Translated from the German.* Edited by T. K. ARNOLD, M.A.

New Edition. 12mo. 9s.

In the Press, New Edition, re-arranged, with fresh Pieces and additional References.

Materials and Models for Greek Prose Composition.

Selected and arranged by J. Y. SARGENT, M.A., *Fellow and Tutor of Magdalen College, Oxford;* and T. F. DALLIN, M.A., *Tutor, late Fellow of Queen's College, Oxford.*

Crown 8vo.

LONDON, OXFORD, AND CAMBRIDGE.

Classical Examination Papers.

Edited, with Notes and References, by P. J. F. GANTILLON, M.A.,
*sometime Scholar of St. John's College, Cambridge; Classical Master
at Cheltenham College.*

Crown 8vo. 7s. 6d.

Or interleaved with writing-paper, half-bound, 10s. 6d.

Recommended in the *Guide to the Choice of Classical Books*, by J. B.
Mayor, M.A., Professor of Classical Literature at King's College, late
Fellow and Tutor of St. John's College, Cambridge.

Demosthenes.

Edited, with English Notes and Grammatical References, by THOMAS
KERCHEVER ARNOLD, M.A.

12mo.

OLYNTHIAC ORATIONS. Third Edition. 3s.
PHILIPPIC ORATIONS. Third Edition. 4s.
ORATION ON THE CROWN. Second Edition. 4s. 6d.

Demosthenis Orationes Privatae.

Edited by ARTHUR HOLMES, M.A., *late Senior Fellow and Dean of
Clare College, Cambridge, and Preacher at the Chapel Royal, Whitehall.*

Crown 8vo.

DE CORONA. 5s.

Forming a Part of the "Catena Classicorum."

Demosthenis Orationes Publicae.

Edited by G. H. HESLOP, M.A., *late Fellow and Assistant-Tutor of
Queen's College, Oxford; Head-Master of St. Bees.*

Crown 8vo.

OLYNTHIACS, 2s. 6d. } or, in One Volume, 4s. 6d.
PHILIPPICS, 3s.
DE FALSA LEGATIONE, 6s.

Forming Parts of the "Catena Classicorum."

Isocratis Orationes.

Edited by JOHN EDWIN SANDYS, M.A., *Fellow and Tutor of
St. John's College, Cambridge.*

Crown 8vo.

AD DEMONICUM ET PANEGYRICUS. 4s. 6d.

Forming a Part of the "Catena Classicorum."

The Greek Testament.

With a Critically Revised Text; a Digest of Various Readings; Marginal References to Verbal and Idiomatic Usage; Prolegomena; and a Critical and Exegetical Commentary. For the use of Theological Students and Ministers.

By HENRY ALFORD, D.D., *late Dean of Canterbury.*

New Edition. 4 vols. 8vo. 102s.

The Volumes are sold separately, as follows:

Vol.　I.—The FOUR GOSPELS. 28s.
Vol.　II.—ACTS to 2 CORINTHIANS. 24s.
Vol. III.—GALATIANS to PHILEMON. 18s.
Vol. IV.—HEBREWS to REVELATION. 32s.

The Greek Testament.

With Notes, Introductions, and Index.

By CHR. WORDSWORTH, D.D., *Bishop of Lincoln.*

New Edition. 2 vols. Impl. 8vo. 60s.

The Parts may be had separately, as follows:—

The GOSPELS. 16s.
The ACTS. 8s.
St. Paul's EPISTLES. 23s.
GENERAL EPISTLES, REVELATION, and INDEX. 16s.

Notes on the Greek Testament.

By the Rev. ARTHUR CARR, M.A., *Assistant-Master at Wellington College, late Fellow of Oriel College, Oxford.*

THE GOSPEL ACCORDING TO S. LUKE.

Crown 8vo. 6s.

(*See Specimen Page, No.* 15.)

Madvig's Syntax of the Greek Language, especially of the Attic Dialect.

For the use of Schools.

Edited by THOMAS KERCHEVER ARNOLD, M.A.

New Edition. Imperial 16mo. 8s. 6d.

LONDON, OXFORD, AND CAMBRIDGE.

Sophocles.

With English Notes from SCHNEIDEWIN.
> *Edited by* T. K. ARNOLD, M.A., ARCHDEACON PAUL, *and* HENRY
BROWNE, M.A.

12mo.

AJAX. 3s.   PHILOCTETES. 3s.   ŒDIPUS TYRANNUS. 4s.   ŒDIPUS
COLONEUS. 4s.   ANTIGONE. 4s.

Sophoclis Tragoediae.

> *Edited by* R. C. JEBB, M.A., *Fellow and Assistant-Tutor of* Trinity
College, Cambridge, *and Public Orator of the University.*

Crown 8vo.

ELECTRA.   Second Edition, revised.   3s. 6d.
AJAX.   3s. 6d.

> *Forming Parts of the "Catena Classicorum."*

Aristophanis Comoediae.

> *Edited by* W. C. GREEN, M.A., *late Fellow of King's College,*
Cambridge; *Assistant-Master at Rugby School.*

Crown 8vo.

THE ACHARNIANS and THE KNIGHTS.   4s.
THE CLOUDS.   3s. 6d.
THE WASPS.   3s. 6d.

An Edition of "THE ACHARNIANS and THE KNIGHTS," revised
and especially prepared for Schools.   4s.

> *Forming Parts of the "Catena Classicorum."*

Herodoti Historia.

> *Edited by* H. G. WOODS, M.A., *Fellow and Tutor of Trinity College,*
Oxford.

Crown 8vo.

BOOK I.   6s.   BOOK II.   5s.
Forming Parts of the "Catena Classicorum."

LONDON, OXFORD, AND CAMBRIDGE.

A Copious Phraseological English-Greek Lexicon.

Founded on a work prepared by J. W. FRÄDERSDORFF, Ph.D., *late Professor of Modern Languages, Queen's College, Belfast.*

Revised, Enlarged, and Improved by the late THOMAS KERCHEVER ARNOLD, M.A., *and* HENRY BROWNE, M.A.

Fourth Edition. 8vo. 21*s.*

Thucydidis Historia. Books I. and II.

Edited by CHARLES BIGG, M.A., *late Senior Student and Tutor of Christ Church, Oxford; Principal of Brighton College.*

Crown 8vo. 6*s.*

Forming a Part of the "Catena Classicorum.

Thucydidis Historia. Books III. and IV.

Edited by G. A. SIMCOX, M.A., *Fellow of Queen's College, Oxford.*

Crown 8vo. 6*s.*

Forming a Part of the "Catena Classicorum."

An Introduction to Aristotle's Ethics.

Books I.—IV. (Book X., c. vi.—ix. in an Appendix). With a Continuous Analysis and Notes. **Intended** for the use of Beginners and Junior Students.

By the Rev. EDWARD MOORE, B.D., *Principal of S. Edmund Hall, and late Fellow and Tutor of Queen's College, Oxford.*

Crown 8vo. 10*s.* 6*d.*

Aristotelis Ethica Nicomachea.

Edidit, emendavit, crebrisque locis parallelis e libro ipso, aliisque ejusdem Auctoris scriptis, illustravit JACOBUS E. T. ROGERS, A.M.

Small 8vo. 4*s.* 6*d.* Interleaved with writing-paper, half-bound. 6*s.*

Selections from Aristotle's Organon.

Edited by JOHN R. MAGRATH, M.A., *Fellow and Tutor of Queen's College, Oxford.*

Crown 8vo. 3*s.* 6*d.*

LONDON, OXFORD, AND CAMBRIDGE.

CATENA CLASSICORUM
Crown 8vo.

Sophoclis Tragoediae.   By R. C. JEBB, M.A.
    THE ELECTRA. 3s. 6d.   THE AJAX. 3s. 6d.

Juvenalis Satirae.   By G. A. SIMCOX, M.A. 5s.

Thucydidis Historia.—Books I. & II.
    By CHARLES BIGG, M.A. 6s.

Thucydidis Historia.—Books III. & IV.
    By G. A. SIMCOX, M.A. 6s.

Demosthenis Orationes Publicae.   By G. H. HESLOP, M.A.
    THE OLYNTHIACS. 2s. 6d.   }
    THE PHILIPPICS. 3s.   or, in One Volume, 4s. 6d.
    DE FALSA LEGATIONE. 6s.

Demosthenis Orationes Privatae.
    By ARTHUR HOLMES, M.A.
    DE CORONA. 5s.

Aristophanis Comoediae.   By W. C. GREEN, M.A.
    THE ACHARNIANS AND THE KNIGHTS. 4s.
    THE WASPS. 3s. 6d.   THE CLOUDS. 3s. 6d.
    An Edition of THE ACHARNIANS and the KNIGHTS, revised and especially adapted
for use in Schools. 4s.

Isocratis Orationes.   By JOHN EDWIN SANDYS, M.A.
    AD DEMONICUM ET PANEGYRICUS. 4s. 6d.

Persii Satirae.   By A. PRETOR, M.A.   3s. 6d.

Homeri Ilias.   By S. H. REYNOLDS, M.A.
    BOOKS I. TO XII. 6s.

Terenti Comoediae.   By T. L. PAPILLON, M.A.
    ANDRIA AND EUNUCHUS. 4s. 6d.
    ANDRIA. New Edition, with Introduction on Prosody. 3s. 6d.

Herodoti Historia.   By H. G. WOODS, M.A.
    BOOK I., 6s.   BOOK II., 5s.

Horati Opera.   By J. M. MARSHALL, M.A.
VOL. I.—THE ODES, CARMEN SECULARE, AND EPODES. 7s. 6d.

Taciti Historiae.   By W. H. SIMCOX, M.A.
    BOOKS I. AND II. 6s.   BOOKS III., IV., and V. [*In the Press.*

DIVINITY

MANUALS OF RELIGIOUS INSTRUCTION

Edited by

JOHN PILKINGTON NORRIS, B.D.,

CANON OF BRISTOL, CHURCH INSPECTOR OF TRAINING COLLEGES.

Each Book in Five Parts.   Small 8vo.   1s. each Part.

Or in Three Volumes.   3s. 6d. each.

"Contain the maximum of requisite information within a surprising minimum of space. They are the best and fullest and simplest compilation we have hitherto examined on the subject treated."

Standard.

"Carefully prepared, and admirably suited for their purpose, they supply an acknowledged want in Primary Schools, and will doubtless be in great demand by the teachers for whom they are intended."

Educational Times.

THE OLD TESTAMENT.

By the Rev. E. J. GREGORY, M.A., *Vicar of Halberton.*

PART I. The Creation to the Exodus.   PART II. Joshua to the Death of Solomon.   PART III. The Kingdoms of Judah and Israel.   PART IV. Hebrew Poetry—The Psalms.   PART V. The Prophets of the Captivity and of the Return—The Maccabees—Messianic Teaching of the Old Testament.

THE NEW TESTAMENT.

By C. T. WINTER.

PART I. St. Matthew's Gospel.   PART II. St. Mark's Gospel. PART III. St. Luke's Gospel.   PART IV. St. John's Gospel.   PART V. The Acts of the Apostles.

THE PRAYER BOOK.

By JOHN PILKINGTON NORRIS, B.D., *Canon of Bristol, &c.*

PART I. The Catechism to the end of the Lord's Prayer—The Order for Morning and Evening Prayer.   PART II. The Catechism, concluding portion—The Office of Holy Baptism—The Order of Confirmation. PART III. The Theology of the Catechism—The Litany—The Office of Holy Communion.   PART IV. The Collects, Epistles, and Gospels, to be used throughout the year.   PART V. The Thirty-Nine Articles.

LONDON, OXFORD, AND CAMBRIDGE.

Rudiments of Theology.

Intended to be a First Book for Students.

By JOHN PILKINGTON NORRIS, B.D., *Canon of Bristol, Church Inspector of Training Colleges.*

Crown 8vo. [*Just Ready.*

A Catechism for Young Children, Preparatory to the Use of the Church Catechism.

By JOHN PILKINGTON NORRIS, B.D., *Canon of Bristol.*

Small 8vo. 2*d.*

A Companion to the Old Testament.

Being a plain Commentary on Scripture History down to the Birth of our Lord.

Small 8vo. 3*s.* 6*d.*

Also in Two Parts :

Part I.—The Creation of the World to the Reign of Saul.
Part II.—The Reign of Saul to the Birth of Our Lord.

Small 8vo. 2*s.* each.

[Especially adapted for use in Training Colleges and Schools.]

" A very compact summary of the Old Testament narrative, put together so as to explain the connection and bearing of its contents, and written in a very good tone ; with a final chapter on the history of the Jews between the Old and New Testaments. It will be found very useful for its purpose. It does not confine itself to merely chronological difficulties, but comments freely upon the religious bearing of the text also."—*Guardian.*

A Companion to the New Testament.

Small 8vo. [*In the Press.*

The Young Churchman's Companion to the Prayer Book.

By the Rev. J. W. GEDGE, M.A., *Diocesan Inspector of Schools for the Archdeaconry of Surrey.*

Part I.—Morning and Evening Prayer and Litany.
Part II.—Baptismal and Confirmation Services.

18mo. 1*s.* each, or in Paper Cover, 6*d.*

Recommended by the late and present LORD BISHOPS OF WINCHESTER.

LONDON, OXFORD, AND CAMBRIDGE.

A Manual of Confirmation.

With a Pastoral Letter instructing Catechumens how to prepare themselves for their First Communion.

By EDWARD MEYRICK GOULBURN, D.D., *Dean of Norwich.*
Ninth Edition. Small 8vo. 1*s.* 6*d.*

The Way of Life.

A Book of Prayers and Instruction for the Young at School. With
a Preparation for Holy Communion.

Compiled by a Priest. Edited by the Rev. T. T. CARTER, M.A.,
Rector of Clewer, Berks.

16mo, 1*s.* 6*d.*

Household Theology.

A Handbook of Religious Information respecting the Holy Bible, the
Prayer Book, the Church, the Ministry, Divine Worship, the Creeds,
&c., &c.

By the Rev. JOHN HENRY BLUNT, M.A.
New Edition. Small 8vo. 3*s.* 6*d.*

Keys to Christian Knowledge.

Small 8vo. 2*s.* 6*d.* each.

" Of cheap and reliable text-books of this nature there has hitherto been a great want. We are often asked to recommend books for use in Church Sunday schools, and we therefore take this opportunity of saying that we know of none more likely to be of service both to teachers and scholars than these *Keys.*" — *Churchman's Shilling Magazine.*

" Will be very useful for the higher classes in Sunday schools, or rather for the fuller instruction of the Sunday-school teachers themselves, where the parish Priest is wise enough to devote a certain time regularly to their preparation for their voluntary task." — *Union Review.*

By J. H. BLUNT, M.A., Editor of the *Annotated Book of Common Prayer.*

THE HOLY BIBLE.

THE BOOK OF COMMON PRAYER.

THE CHURCH CATECHISM.

CHURCH HISTORY, ANCIENT.

CHURCH HISTORY, MODERN.

By JOHN PILKINGTON NORRIS, B.D., *Canon of Bristol.*

THE FOUR GOSPELS.

THE ACTS OF THE APOSTLES.

LONDON, OXFORD, AND CAMBRIDGE.

MISCELLANEOUS

A First German Accidence and Exercise Book.

By J. W. J. VECQUERAY, *Assistant-Master at Rugby School.*

[*In preparation.*

Selections from La Fontaine's Fables.

Edited, with English Notes at the end, for use in Schools, by P. BOWDEN-SMITH, M.A., *Assistant-Master at Rugby School.*

[*In preparation.*

Le Maréchal de Villars, from Ste. Beuve's Causeries du Lundi.

Edited, with English Notes at the end, for use in Schools, by H. W. EVE, M.A., *Assistant-Master at Wellington College, sometime Fellow of Trinity College, Cambridge.*

[*In preparation.*

The Campaigns of Napoleon.

The Text (in French) from M. THIERS' *"Histoire du Consulat et de l'Empire," and "Histoire de la Révolution Française." Edited, with English Notes, for the use of Schools, by* EDWARD E. BOWEN, M.A., *Master of the Modern Side, Harrow School.*

With Maps.   Crown 8vo.

ARCOLA. 4s. 6d.   MARENGO. 4s. 6d.
JENA. 3s. 6d.   WATERLOO. 6s.

Selections from Modern French Authors.

Edited, with English Notes and Introductory Notice, by HENRI VAN LAUN, *Translator of Taine's* HISTORY OF ENGLISH LITERATURE.

Crown 8vo.   3s. 6d. each.

HONORÉ DE BALZAC.   H. A. TAINE.

LONDON, OXFORD, AND CAMBRIDGE.

The First French Book.
By T. K. ARNOLD, M.A.
Sixth Edition. 12mo. 5s. 6d. Key, 2s. 6d.

The First German Book.
By T. K. ARNOLD, M.A., *and* J. W. FRÄDERSDORFF, Ph.D.
Seventh Edition. 12mo. 5s. 6d. Key, 2s. 6d.

The First Hebrew Book.
By T. K. ARNOLD, M.A.
Fourth Edition. 12mo. 7s. 6d. Key, 3s. 6d.

The Choristers' Guide.
By W. A. BARRETT, Mus. Bac., Oxon., *of St. Paul's Cathedral,*
Author of "Flowers and Festivals," &c.
Second Edition. Square 16mo. 2s. 6d.

Form and Instrumentation.
By W. A. BARRETT, Mus. Bac., Oxon., *Author of "The Choristers'*
Guide," &c.
Small 8vo. [*In preparation.*

these too far apart, and the intercourse of the defenders with an army
of relief under the Count of Clermont at Blois was not broken off.
Early in the following year, this army hoped to raise the siege by
falling on a large body of provisions coming to the besiegers from
Battle of the Paris under Sir John Fastolf. The attack was made at
Herrings. Rouvray, but Fastolf had made careful preparations.
The waggons were arranged in a square, and, with the stakes of the
archers, formed a fortification on which the disorderly attack of the
French made but little impression. Broken in the assault, they fell
an easy prey to the English, as they advanced beyond their lines.
The skirmish is known by the name of the Battle of the Herrings.
This victory, which deprived the besieged of hope of external succour,
seemed to render the capture of the city certain.

Already at the French King's court at Chinon there was talk of a
Danger of hasty withdrawal to Dauphiné, Spain, or even Scotland;
Orleans. when suddenly there arose one of those strange effects
of enthusiasm which sometimes set all calculation at defiance.

In Domrémi, a village belonging to the duchy of Bar, the inhabi-
tants of which, though in the midst of Lorraine, a province under
Burgundian influence, were of patriotic views, lived a village maiden
called Joan of Arc. The period was one of great mental excitement;
as in other times of wide prevailing misery, prophecies and mystical
preachings were current. Joan of Arc's mind was particularly
susceptible to such influences, and from the time she
Joan of Arc. was thirteen years old, she had fancied that she heard
voices, and had even seen forms, sometimes of the Archangel Michael,
sometimes of St. Catherine and St. Margaret, who called her to
the assistance of the Dauphin. She persuaded herself that she was des-
tined to fulfil an old prophecy which said that the kingdom, destroyed
by a woman—meaning, as she thought, Queen Isabella,—should be
saved by a maiden of Lorraine. The burning of Domrémi in the
summer of 1428 by a troop of Burgundians at length gave a practical
form to her imaginations, and early in the following year she suc-
ceeded in persuading Robert of Baudricourt to send her, armed and
accompanied by a herald, to Chinon. She there, as it is said by the
wonderful knowledge she displayed, convinced the court of the truth
of her mission. At all events, it was thought wise to take advantage
of the infectious enthusiasm she displayed, and in April she was
intrusted with an army of 6000 or 7000 men, which was to march up
the river from Blois to the relief of Orleans. When she appeared
upon the scene of war, she supplied exactly that element of success

of all of them open by two slits turned towards the centre of the flower. Their stalks have expanded and joined together, so as to form a thin sheath round the central column (fig. 12). The dust-

Fig. 12.
Dust-spikes of gorse (*enlarged*).

spikes are so variable in length in this flower, that it may not be possible to see that one short one comes between two long ones, though this ought to be the case.

The *seed-organ* is in the form of a longish rounded pod, with a curved neck, stretching out beyond the dust-spikes. The top of it is sticky, and if you look at a bush of gorse, you will see it projecting beyond the keel in most of the fully-blown flowers, because the neck has become more curved than in fig. 12. Cut open the pod; it contains only one cavity (not, as that of the wall-flower, two separated by a thin partition), and the grains are suspended by short cords from the top (fig. 13). These grains may be plainly seen in the seed-organ of even a young flower. It is evident that they are the most important part of the plant, as upon them depends its diffu-

Fig. 13.
Split seed-pod of gorse.

sion and multiplication. We have already seen how carefully their well-being is considered in the matter of their perfection, how even insects are pressed into their service for this purpose! Now let us glance again at our flower, and see how wonderfully contrivance is heaped upon contrivance for their protection !

First (see fig. 10, p. 14), we have the outer covering, so covered with hairs, that it is as good for keeping out rain as a waterproof cloak; in the buttercup, when you pressed the bud, it separated into five leaves; here there are five leaves, just the same, but they are so tightly joined that you may press till the whole bud is bent without making them separate at all, and when the bud is older, they only separate into two, and continue to enfold the flower to a certain extent till it fades. When the flower pushes back its waterproof cloak, it has the additional shelter of the big

struction, and at last, after nearly twenty years of alternate
hopes and fears, of tedious negotiations, official evasions,
and sterile Parliamentary debates, it was effectually extin-
guished by the adverse report of a Parliamentary Committee, followed by the erection of the present Millbank
Penitentiary at a vastly greater expense and on a totally
different system.

Transportation.—In the meantime the common gaols
were relieved in a makeshift fashion by working gangs of
prisoners in hulks at the seaports; but the resource mainly
relied on for getting rid of more dangerous criminals was
the old one of transportation, Botany Bay having suc-
ceeded to America. As at first employed, there was no
mistake as to the reality of the punishment; the mis-
fortune was that the worst elements in the real were not
so made known as to form any part of the apparent
punishment. If the judge, in sentencing the convict,
had thought fit to explain, for the warning of would-be
offenders, exactly what was going to be done with their
associate, the sentence would have been something of
this sort: " You shall first be kept, for days or months
as it may happen, in a common gaol, or in the hulks, in
company with other criminals better or worse than your-
self, with nothing to do, and every facility for mutual
instruction in wickedness. You shall then be taken on
board ship with similar associates of both sexes, crammed
down between decks, under such circumstances that
about one in ten of you will probably die in the course
of the six months' voyage. If you survive the voyage
you will either be employed as a slave in some public
works, or let out as a slave to some of the few free
settlers whom we have induced to go out there. In
either case you will be under very little regular inspection,
and will have every opportunity of indulging those natural

wealth into the **treasury**. Churches remained open day and night, and frequent addresses kept up the enthusiasm to a high pitch. It was (for the moment) a genuine "revival" or reawakening of the whole Roman world. The occasion, too, appeared favourable. Italy was quiet, and the Exarchate at peace with its neighbours. Clotaire the Frank was no enemy to Heraclius, and in common with his clergy (being orthodox and not Arian) might be expected to sympathise in so holy a cause.

Treachery of the Avars—A.D. 616.—In one quarter only was there room for fear. The Avars were on the Danube, and the turbulence of the Avars was only equalled by their perfidy. Already, in A.D. 610, they had fallen suddenly on North Italy, and pillaged and harassed those same Lombards whom they had before helped to destroy the Gepidæ. Previous to an absence, therefore, of years from his capital, it was essential for the Emperor to sound their intentions, and, if possible, to secure their neutrality. His ambassadors were welcomed with apparent cordiality, and an interview was arranged between the Chagan and Heraclius. The place was to be Heraclea. At the appointed time the Emperor set out from Selymbria to meet the Khan, decked with Imperial crown and mantle to honour the occasion. The escort was a handful of soldiers; but there was an immense cortége of high officials and of the fashionable world of Constantinople, and the whole country side was there to see. Presently some terrified peasants were seen making their way hurriedly towards Heraclius. They urged him to flee for his life; for armed Avars had been seen in small bodies, and might even now be between him and the capital. Heraclius knew too much to hesitate. He threw off his robes and fled, and but just in time. The Chagan had laid a deep plot. A large mass of men had been told off in small detachments

I say the pulpit (in the sober use
Of its legitimate peculiar pow'rs)
Must stand acknowledg'd, while the world shall stand,
The most **important** and effectual guard,
Support and ornament of virtue's cause.
There stands the messenger of truth : there stands
The legate of the skies; his theme divine,
His office sacred, his credentials clear.
By him, the violated law speaks out   340
Its thunders, and by him, in strains as sweet
As angels use, the Gospel whispers peace.
He stablishes the strong, restores the weak,
Reclaims the wand'rer, binds the broken heart,
And, arm'd himself in panoply complete
Of heav'nly temper, furnishes with arms
Bright as his own, and trains, by ev'ry rule
Of holy discipline, to glorious war,
The sacramental host of God's elect.
Are all such teachers? would to heav'n all were!   350
But hark—the Doctor's voice—fast wedged between
Two empirics he stands, and with swoln cheeks
Inspires the news, his trumpet. Keener far
Than all invective is his bold harangue,
While through that public organ of report
He hails the clergy; and, defying shame,
Announces to the world his own and theirs.
He teaches those to read, whom schools dismiss'd,
And colleges, untaught; sells accent, tone,
And emphasis in score, and gives to pray'r   360
Th' *adagio* and *andante* it demands.
He grinds divinity of other days
Down into modern use; transforms old **print**
To zigzag manuscript, and cheats the eyes
Of gall'ry critics by a thousand arts.—
Are there who purchase of the Doctor's ware?
Oh name it not in Gath !—it cannot be,
That grave and learned Clerks should need such aid.
He doubtless is in sport, and does but droll,
Assuming thus a rank unknown before,   370
Grand caterer and dry-nurse of the church.

I venerate the man whose heart is warm,
Whose hands are pure. whose doctrine and whose life.

gether as with a close seal. . . . . The flakes of his flesh are joined together: they are firm in themselves; they cannot be moved."

Hobbes, in his famous book to which he gave the title *Leviathan,* symbolised thereby the force of civil society, which he made the foundation of all right.

315–325 Cowper's limitation of the province of satire—that it is fitted to laugh at foibles, not to subdue vices—is on the whole well-founded. But we cannot forget Juvenal's famous "facit indignatio versum," or Pope's no less famous—

> "Yes, I am proud: I must be proud to see
> Men not afraid of God, afraid of me:
> Safe from the bar, the pulpit, and the throne,
> Yet touched and shamed by ridicule alone."

326–372 *The pulpit, not satire, is the proper corrector of sin. A description of the true preacher and his office, followed by one of the false preacher, "the reverend advertiser of engraved sermons."*

330 *Strutting and vapouring.* Cf. *Macbeth,* v. 5.

> "Life's but a walking shadow, a poor player,
> That struts and frets his hour upon the stage,
> And then is heard no more; it is a tale
> Told by an idiot, full of sound and fury,
> Signifying nothing."

> "And what in real value's wanting,
> Supply with vapouring and ranting."—HUDIBRAS.

331 *Proselyte.* προσήλυτος, a new comer, a convert to Judaism.

338 *His theme divine.* Nominative absolute.

343 *Stablishes.* Notice the complete revolution the word has made—stabilire, établir, establish, stablish; cf. state, &c.

346 *Of heavenly temper.* Cf. *Par. Lost,* i. 284, "his ponderous shield etherial temper." See note on *Winter Morning Walk,* l. 664.

349 *Sacramental.* Used in the Latin sense. Sacramentum was the oath of allegiance of a Roman soldier. The word in its Christian sense was first applied to baptism—the vow to serve faithfully under the banner of the cross. See *Browne on the Thirty-nine Articles,* p. 576.

350 *Would to heaven.* A confusion between "would God" and "I pray to heaven."

351 A picture from the life of a certain Dr Trusler, who seems to have combined the trades of preacher, teacher of elocution, writer of sermons, and literary hack.

352 *Empirics.* ἐμπειρικός, one who trusts solely to experience or practice instead of rule, hence a quack. The accent is the same as in Milton (an exception to the rule. See note on *Sofa,* l. 52).

thus : if the articles had cost £1 each, the total cost
would have been £2478 ;

∴ as they cost ⅙ of £1 each. the cost will be £$\frac{2478}{6}$, or £413.

The process may be written thus :

3s. 4d. is ⅙ of £1 | £2478 = cost of the articles at £1 each.

£413 = cost .... at 3s. 4d. ...

Ex. (2). Find the cost of 2897 articles at £2. 12s. 0d.
each.

£2 is 2 × £1	2897 . 0 . 0 = cost at £1 each.
10s. is ½ of £1	5794 . 0 . 0 = ... £2 ....
2s. is ⅕ of 10s.	1448 . 10 . 0 = ... 10s.....
8d. is ⅓ of 2s.	289 . 14 . 0 = ... 2s. ....
1d. is ⅛ of 8d.	96 . 11 . 4 = ... 8d. ....
	12 . 1 . 5 = ... 1d. ....

£7640 . 16 . 9 = ... £2. 12s. 9d. each.

NOTE.—A shorter method would be to take the parts
thus :

10s. = ½ of £1 ; 2s. 6d. = ¼ of 10s. ; 3d. = $\frac{1}{10}$ of 2s. 6d.

Ex. (3). Find the cost of 425 articles at £2. 18s. 4d.
each.

Since £2. 18s. 4d. is the difference between £3 and
1s. 8d. (which is $\frac{1}{12}$ of £1), the shortest course is to find
the cost at £3 each, and to *subtract from it* the cost at
1s. 8d. each, thus :

	£ s. d.
£3 is 3 × £1	425 . 0 . 0 = cost at £1 each.
1s. 8d. is $\frac{1}{12}$ of £1	1275 . 0 . 0 = ... £3 ....
	35 . 8 . 4 = ... 1s. 8d. each.

£1239 . 11 . 8 = ... £2. 18s. 4d. each.

[J. HAMBLIN SMITH's ARITHMETIC— *See Page* 10.]

PROPOSITION XLI. THEOREM.

If a parallelogram and a triangle be upon the same base, and between the same parallels, the parallelogram is double of the triangle.

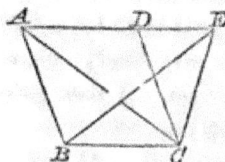

Let the $\square$ *ABCD* and the $\triangle$ *EBC* be on the same base *BC* and between the same ‖s *AE, BC*.

Then must $\square$ *ABCD be double of* $\triangle$ *EBC.*

Join *AC*.

Then $\triangle ABC = \triangle EBC$, $\because$ they are on the same base and between the same ‖s ;   I. 37.

and $\square$ *ABCD* is double of $\triangle ABC$, $\because$ *AC* is a diagonal of *ABCD* ;   I. 34.

        $\therefore$ $\square$ *ABCD* is double of $\triangle EBC$.

                                Q. E. D.

Ex. 1. If from a point, without a parallelogram, there be drawn two straight lines to the extremities of the two opposite sides, between which, when produced, the point does not lie, the difference of the triangles thus formed is equal to half the parallelogram.

Ex. 2. The two triangles, formed by drawing straight lines from any point within a parallelogram to the extremities of its opposite sides, are together half of the parallelogram.

[J. HAMBLIN SMITH'S GEOMETRY—*See Page* II.]

Sometimes carbonic anhydride is produced in wells, and, being so much heavier than air, it remains at the bottom. If a man goes down into such a well, he will have no difficulty at first, because the air is good; but when he is near the bottom, where the gas has accumulated, he will gasp for breath and fall; and if anyone, not understanding the cause of his trouble, goes down to assist him, he too will fall senseless, and both will quickly die. The way to ascertain whether carbonic anhydride has accumulated at the bottom of a well is to let a light down into it. If it goes out, or even burns very dimly, there is enough of the gas to make the descent perilous. A man going down a well should always take a candle with him, which he should hold a considerable distance below his mouth. If the light burns dimly, he should at once stop, before his mouth gets any lower and he takes some of the gas into his lungs.

When this gas is in a well or pit, of course it must be expelled before a man can descend. There are several expedients for doing this. One is to let a bucket down frequently, turning it upside down, away from the mouth of the well, every time it is brought up, a plan which will remind you of the experiment represented in Fig. 24.

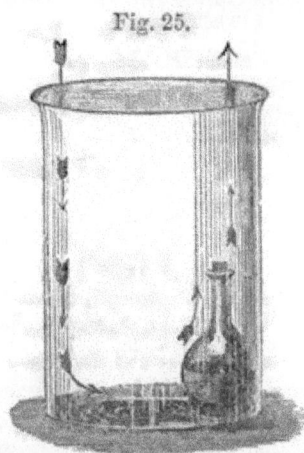

Fig. 25.

But a better way is to let down a bundle of burning straw or shavings, so as to heat the gas. Now heated bodies expand, gases very much more than solids or liquids, and, in expanding, the weight of a certain volume, say of a gallon, becomes lessened. So that if we can heat the carbonic anhydride enough to make a gallon of it weigh less than a gallon of air, it will rise out of the well just as hydrogen gas would do. Fig. 25 shows how you may perform this experiment upon a small scale.

DISASTROUS RETREAT OF THE ENGLISH FROM CABUL.

IT took two days of disorder, suffering, and death to carry the army, now an army no more, to the jaws of the fatal pass. Akbar Khan, who appeared like the Greeks' dread marshal from the spirit-land at intervals upon the route, here demanded four fresh hostages. The demand was acquiesced in. Madly along the narrow defile crowded the undistinguishable host, whose diminished numbers were still too numerous for speed : on every side rang the war-cry of the barbarians : on every side plundered and butchered the mountaineers : on every side, palsied with fatigue, terror, and cold, the soldiers dropped down to rise no more. The next day, in spite of all remonstrance, the general halted his army, expecting in vain provisions from Akbar Khan. That day the ladies, the children, and the married officers were given up. The march was resumed. By the following night not more than one-fourth of the original number survived. Even the haste which might once have saved now added nothing to the chances of life. In the middle of the pass a barrier was prepared. There twelve officers died sword in hand. A handful of the bravest or the strongest only reached the further side alive : as men hurry for life, they hurried on their way, but were surrounded and cut to pieces, all save a few that had yet escaped. Six officers better mounted or more fortunate than the rest, reached a spot within sixteen miles of the goal ; but into the town itself rode painfully on a jaded steed, with the stump of a broken sword in his hand, but one.

LIVY, xxi. c. 25, § 7-10. xxxv. c. 30. xxiii. c. 24.
CÆSAR, Bell. Gall. v. c. 35-37.

DEFEAT OF CHARLES THE BOLD AND MASSACRE OF HIS TROOPS AT MORAT.

IN such a predicament braver soldiers might well have ceased to struggle. The poor wretches, Italians and Savoyards, six thousand or more in number, threw away their arms and made

II.

ARIADNE'S LAMENT.

Madam, 'twas Ariadne passioning
For Theseus' perjury and unjust flight.
TWO GENTLEMEN OF VERONA, IV. 4, 172.

ARGUMENT.

ARIADNE *tells the story of her first waking, to find herself abandoned by Theseus and left on an unknown island, exposed to a host of dangers.*—(HEROIDES, X.)

The story is beautifully told by Catullus, in the "Epithalamium Pelei et Thetidos:" it also forms one of the episodes in Chaucer's "Legende of Goode Women."

I woke before it was day to find myself alone, no trace of my companions to be seen. In vain I felt and called for Theseus; the echoes alone gave me answer.

QUAE legis, ex illo, Theseu, tibi litore mitto,
 Unde tuam sine me vela tulere ratem :
In quo me somnusque meus male prodidit et tu,
 Per facinus somnis insidiate meis.   107
Tempus erat, vitrea quo primum terra pruina   112
 Spargitur et tectae fronde queruntur aves :
Incertum vigilans, a somno languida, movi   97
 Thesea prensuras semisupina manus :
Nullus erat, referoque manus, iterumque retempto,
10 Perque torum moveo brachia : nullus erat.
Excussere metus somnum : conterrita surgo,
 Membraque sunt viduo praecipitata toro.   123
Protinus adductis sonuerunt pectora palmis,   111
 Utque erat e somno turbida, rapta coma est.
Luna fuit : specto, siquid nisi litora cernam ;
 Quod videant, oculi nil nisi litus habent.   150
Nunc huc, nunc illuc, et utroque sine ordine curro ;
 Alta puellares tardat arena pedes.
Interea toto clamanti litore " Theseu !"   121
20 Reddebant nomen concava saxa tuum,
Et quoties ego te, toties locus ipse vocabat :
 Ipse locus miserae ferre volebat opem.   106₃

STORIES FROM OVID.

174. **Punica poma,** pomegranates.
178. **Taenarum,** at the southern extremity of Peloponnesus, was one
     of the numerous descents to Tartarus. Cf. Virgil, Georg.
     IV. 467 :

     > Taenarias etiam fauces, alta ostia Ditis.

179. **Factura fuit.** This periphrasis for *fecisset* is to be noted ; it is
     the one from which the oblique forms are all constructed,
     e.g., facturam fuisse, or *factura fuisset.*
183. **Cessatis,** one of a goodly number of intransitive verbs of the
     first conjugation which have a passive participle. Cf. **erratas,**
     above, 139, **clamata,** 35. So Horace, regnata Phalanto rura
     (Odes, II. 6, **12**) ; triumphatae gentes (Virgil).

II.—IV.

ARIADNE.

THIS and the two following extracts, though taken from different
works, form a definite sequence. Ariadne, daughter of Minos, king of
Crete, has helped Theseus to conquer the Minotaur, by giving him a
clew to the maze in which the monster was hid, and, being in love with
him, has fled in his company. They put in for the night to the island of
Dia, and Theseus on the next morning treacherously sails away, leaving
the poor girl alone. The first extract is part of an epistle which she is
supposed to write on the day when she discovers his perfidy.
The name Dia, which belonged properly to a small island off the
north coast of Crete, was also a poetical name for Naxos, one of the
largest of the Cyclades. It may have been this fact which led to the
further legend which is recounted in the next extract, how Ariadne,
lorn of Theseus, becomes the bride of Bacchus ; for Naxos was the
home of the Bacchic worship. As the completion of the legend she is
raised to share in Bacchus' divine honours, and as the Cretan Crown
becomes one of the signs of the heavens.

II.

ARIADNE'S LAMENT.

1. **Illo,** sc. *Diae.*
4. **Per facinus,** criminally.
5. Describing apparently the early dawn, or the hour that precedes
   it, when the night is at its coldest, and the birds, half-awake,
   begin to stir in their nests. **Pruina** hints that it is autumn.
7. A beautifully descriptive line—But half-awake, with all the
   languor of sleep still on me.
   A somno = after, as the *result* of.
8. **Semisupina,** on my side, lit., half on my back, describes the
   motion of a person thus groping about on waking. Cf.
   Chaucer :

   > Ryght in the dawenynge awaketh shee,
   > And gropeth in the bed, and fonde ryghte noghte.

55 haec mea magna **fides**? at non, Euandre, pudendis
    volneribus pulsum aspicies, **nec** sospite dirum
    optabis nato funus pater. ei mihi, quantum
    praesidium Ausonia, et quantum tu perdis, Iule !
        Haec ubi deflevit, tolli miserabile corpus
60 imperat, **et** toto lectos ex agmine mittit
    mille viros, qui supremum comitentur honorem,
    intersintque patris lacrimis, solacia luctus
    exigua ingentis, misero set debita **patri.**
    haut segnes alii crates et molle **feretrum**
65 arbuteis texunt virgis et vimine **querno,**
    extructosque toros obtentu frondis inumbrant.
    hic iuvenem agresti sublimem stramine ponunt ;
    qualem virgineo demessum pollice florem
    seu mollis violae, seu languentis hyacinthi,
70 cui neque fulgor adhuc, nec dum sua forma recessit ;
    non iam mater alit tellus, viresque ministrat.
    tunc geminas vestes auroque ostroque rigentis
    extulit Aeneas, quas illi laeta laborum
    ipsa suis quondam manibus Sidonia Dido
75 fecerat, et tenui **telas** discreverat auro.
    harum unam iuveni **supremum maestus** honorem
    induit, arsurasque comas obnubit amictu ;
    multaque **praeterea Laurentis** praemia pugnae
    aggerat, et longo praedam iubet ordine duci.
80 addit equos et tela, quibus spoliaverat hostem.
    vinxerat et post terga manus, quos mitteret umbris
    inferias, **caeso sparsuros** sanguine flammam ;
    indutosque iubet truncos hostilibus armis
    ipsos ferre duces, inimicaque nomina figi.
85 ducitur infelix aevo confectus Acoetes,
    pectora nunc foedans **pugnis,** nunc unguibus ora ;
    sternitur et toto proiectus **corpore** terrae.

Comp. *Geor.* ii. 80, *Nec longum tempus et . . . exiit . . . arbos,* C.
But as these are the only two instances of the construction adduced it is
perhaps safer to take *et* = even.

51 nil iam, etc.] The father is making vows to heaven in his son's
behalf, but the son is gone where vows are neither made nor paid.

55 haec mea magna **fides]** 'Is this the end of all my promises?'
Magna may be taken as 'solemn,' or 'boastful.'

pudendis volneribus] All his wounds are on his breast.

56 dirum optabis **funus** = *morti devovebis.* Compare the meaning of
dirae, xii. 845.

59-99] A description of the funeral rites. Aeneas bids his last farewell.

59 Haec ubi deflevit] 'His moan thus made.' *De* in composition has
two opposite meanings : (1) cessation from or removal of the fundamental
ideas, as in *decresco, dedoceo,* etc.; (2) (as here) in intensifying, as *debello,
demiror, desaevio.*

61 honorem] *Honos* is used by V. for (1) a sacrifice, iii. 118; (2) a
hymn, *Geor.* ii. 393; (3) beauty, *Aen.* x. 24; (4) the 'leafy honours' of
trees, *Geor.* ii. 404; (5) funeral rites, vi. 333, and here. See below, *l.* 76.

63 solatia] In apposition to the whole sentence ; whether it is nom. or
acc. depends on how we resolve the principal sentence ; here, though
solatia applies to the whole sentence, its construction probably depends on
the last clause, which we may paraphrase, *ut praesentes* (τὸ μετεῖναι) *sint
solatia ;* therefore it is nom.

64 crates et molle feretrum] The bier of pliant osier : cf. *l.* 22.

66] Cf. Statius, *Theb.* vi. 55, *torus* et puerile feretrum.

obtentu frondis] 'A leafy canopy.' C. understands 'a layer of leaves.'

67 agresti stramine] 'The rude litter.'

68] Cf. ix. 435 ; *Il.* viii. 306,

μήκων δ' ὡς ἑτέρωσε κάρη βάλεν, ἥτ ἐνὶ κήπῳ
καρπῷ βριθομένη νοτίῃσί τε εἰαρινῇσιν·
ὡς ἑτέρωσ' ἤμυσε κάρη πήληκι βαρυνθέν.

'Even as a flower,
Poppy or hyacinth, on its broken stem
Languidly raises its encumbered head.'—MILMAN.

69 languentis hyacinthi] The rhythm is Greek. The 'drooping hyacinth' is probably the Lilium Martagon or Turk's-cap lily, 'the sanguine
flower inscribed with woe.'

70] 'That hath not yet lost its gloss nor all its native loveliness.' *Recessit* must apply to both clauses. 'If we suppose the two parts of the
line to contain a contrast, the following line will lose much of its force,'
C. Compare the well-known lines from the *Giaour,* 'He who hath bent
him o'er the dead,' etc.

71] Contrast the force of *neque adhuc, nec dum,* and *non iam ;* 'the
brightness not all gone,' 'the lines where beauty lingers,' and 'the support
and nurture of mother earth cut off once and for all.'

36. ἵνα φάγῃ] In modern Greek, which properly speaking has no infinitive, the sense of the infinitive is expressed by νά (ἵνα) with subjunctive (as in this passage), *e.g.* ἐπιθυμῶ νὰ γράφῃ, 'I wish him to write;' see Corfe's *Modern Greek Grammar*, p. 78. This extension of the force of ἵνα to oblique petition, and even to consecutive clauses, may be partly due to the influence of the Latin *ut;* cf. ch. xvi. 27, ἐρωτῶ οὖν, πάτερ, ἵνα πέμψῃς : see note on ch. iv. 3.

The following incident is recorded by St. Luke alone. Simon the Pharisee is not to be identified with Simon the leper, Matt. xxvi., Mark xiv. 3.

ἀνεκλίθη] The Jews had adopted the Roman, or rather Greek, fashion of reclining at meals—a sign of advancing luxury and of Hellenism, in which however even the Pharisee acquiesces.

37. γυνή] There is no proof that this woman was Mary Magdalene. But mediæval art has identified the two, and great pictures have almost disarmed argument in this as in other incidents of the gospel narrative.

38. ἀλάβαστρον] The neuter sing. is Hellenistic. The classical form is ἀλάβαστρος with a heteroclite plural ἀλάβαστρα, hence probably the late sing. ἀλάβαστρον. The grammarian stage of a language loves uniformity, Herod. iii. 20; Theocr. xv. 114 :

Συρίω δὲ μύρω χρύσει' ἀλάβαστρα.

στᾶσα παρὰ τοὺς πόδας αὐτοῦ] This would be possible from the arrangement of the triclinium.

39. ἐγίνωσκεν ἄν] 'Would (all the while) have been recognising.'

40. χρεωφειλέται] A late word; the form varies between χρεωφειλέται and χρεοφειλέται.

41. δηνάρια] The denarius was a silver coin originally containing ten ases (deni), afterwards, when the weight of the as was reduced, sixteen ases. Its equivalent modern value is reckoned at 7½d. But such calculations are misleading; it is more to the point to regard the denarius as an average day's pay for a labourer.

42. μὴ ἐχόντων] Because *he saw that* they had not.
ἐχαρίσατο] Cf. *v.* 21.

INDEX

www.ingramcontent.com/pod-product-compliance
Lightning Source LLC
Chambersburg PA
CBHW021518210326
41599CB00012B/1295